THE
AMERICAN STANDARD
OF PERFECTION

ILLUSTRATED

A COMPLETE DESCRIPTION OF ALL
RECOGNIZED VARIETIES OF FOWLS

AS REVISED BY
THE AMERICAN POULTRY ASSOCIATION
AT ITS THIRTY-EIGHTH ANNUAL MEETING AT ATLANTIC
CITY, N. J., NINETEEN HUNDRED THIRTEEN, AND AT
ITS THIRTY-NINTH ANNUAL MEETING AT CHICAGO,
ILL., NINETEEN HUNDRED FOURTEEN

PRINTED AND PUBLISHED BY
THE AMERICAN POULTRY ASSOCIATION
1915

TO WHOM IT MAY CONCERN:

The public is expressly forbidden, on penalty of the law, to reproduce, duplicate, copy, seek to imitate or to make any other improper use of any of the illustrations contained in this book, all of which are the exclusive property of The American Poultry Association, and protected by copyright in the United States, England and Canada. Permission to make quotations from the text of this book is granted, provided such quotations are disconnected, few in number, and are used solely for the dissemination of knowledge; but these quotations must not be used to an extent nor in a manner that will injure the sale of this work, nor may they be used for advertising purposes, as in circulars, catalogues, etc.

Notice is hereby given that any infringement of the copyright on the contents of this book will result in immediate prosecution.

THE AMERICAN POULTRY ASSOCIATION.

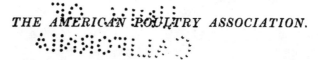

GENERAL INTRODUCTION.

The organization of the American Poultry Association was effected at Buffalo, New York, February, 1873, by delegates from different state and county associations, prominent breeders, fanciers, and other interested persons from different sections of the United States and Canada. Mr. W. H. Churchman of Wilmington, Delaware, was the first president and Mr. J. M. Wade of Philadelphia, the first secretary.

At that time the fundamental object of this organization was to standardize the different varieties of domestic and ornamental fowls, and to that end, a complete Standard of Excellence for all varieties then recognized, was formulated and adopted which was recommended as the guide for judging at all poultry exhibitions. The American Poultry Association has since broadened its scope. Its annual conventions have visited nearly all of our large industrial centers.

The first edition of the Standard was issued in February, 1874. It has been followed by several revised editions but the work of the first Standard makers was so thorough, accurate and far-seeing that but few changes, and those of minor importance, have been necessary. Many new breeds and varieties, nearly all of later origin, have been admitted. After a few editions, the title "Standard of Excellence" was changed to read "Standard of Perfection" as one, theoretically at least, more in accord with its prescribed ideals.

Until 1905, all editions contained text descriptions only, and no attempt was made to delineate ideal fowls. The 1905 edition contains this innovation. The illustrations were line drawings by the best known poultry artists of that time. These were received with approval, in sufficient measure, so

3

365653

that the plan of presenting outline illustrations of many of the leading varieties was continued. The type of illustrations was, however, changed to half-tone illustrations of retouched and idealized photographs of living specimens. These appeared in the 1910 edition after having been approved by the Thirty-fifth Annual Convention.

INTRODUCTION TO THE 1915 EDITION.

It has been the general policy of the American Poultry Association to revise the Standard of Perfection every five years. In deference to this policy and in compliance with the constitutional requirement, a standard Revision Committee of seven, representing as far as possible the interests of all sections of the country and of the different breed classifications, was appointed by the President at the thirty-sixth annual meeting at Denver, Colorado, August 9th, 1911, to recommend such changes in the text and illustrations of the 1910 edition as seemed advisable. The first meeting of this committee · was held at Indianapolis in 1913.

In order to do the most efficient work the committee was divided into several sub-committees. According to a previous plan, the Chairman of each sub-committee had been in touch with the officials of the corresponding specialty clubs and with the most experienced and enthusiastic breeders of the varieties under consideration. Conferences with those interested, particularly with those who were advocating the admission of new breeds or varieties, were also sought.

A complete report was adopted at this meeting, subject to such changes and additions as might seem advisable before the thirty-eighth annual meeting. Further hearings were held at Atlantic City prior to the opening of the annual meeting. The report presented was, after some amendment, adopted by the Association at the thirty-eighth annual meeting at Atlantic City, August 13th, 14th and 15th, 1913.

The committee elected four of its members to serve with the Chairman as a committee to edit and publish the 1915

5

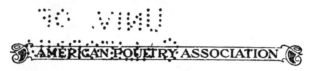

edition. This action was ratified by the executive committee and Association. The committee was given power to contract with artists for new illustrations or for the revision of old ones and instructed to report at the thirty-ninth annual meeting.

In the interval intervening, the committee held further conferences with breeders and fanciers representing different interests and with the artists engaged to prepare the illustrations, which it was decided should be retouched and idealized photographs of living models. The final report was presented to the thirty-ninth annual meeting in Chicago, August 11th, 12th, 13th and 14th, 1914. All the illustrations as far as completed were submitted to the members of the Association present for criticism. The report with few amendments and minor changes voted in the illustrations was adopted by the Association.

The 1915 edition of the Standard of Perfection as herein presented embodied all changes, recommendations, and instructions of the thirty-eighth and thirty-ninth annual meetings, both as to text and illustrations.

TO THE POULTRY ASSOCIATIONS OF AMERICA

RECOGNIZING your loyalty to The American Poultry Association, and believing that a few general rules will be beneficial in conducting poultry exhibitions at which THE AMERICAN STANDARD OF PERFECTION is advertised to govern the placing of the awards, we submit the following, with the request that the same importance be attached to them as to other parts of the STANDARD:

Poultry associations at whose exhibitions THE AMERICAN STANDARD OF PERFECTION is used, are requested to give preference to judges who are members of The American Poultry Association. Judges who are members of this organization are accredited thereby, to the extent of being in good standing, and it is reasonable to conclude that they will apply properly the law of the Association as contained in its STANDARD.

Judges employed by you should be required to follow and apply the STANDARD literally, carefully considering each section of every specimen, according to the scale of points provided for the several breeds. No section is to be ignored. Each section is regarded as important by The American Poultry Association, and should a judge pass a specimen without considering all points, the exhibitor shall be allowed the privilege of protesting the decision, and such protest is to be entertained and properly disposed of by the local association.

Protests are to be entertained by local associations only in cases of apparent dishonesty, ignorance or carelessness on the part of the judge. In scoring the specimens in dispute, the judge, together with the president and secretary of the local association (or representatives appointed by the management of the local association), shall constitute a committee of three, and the majority decision of this committee shall be final. Score cards made out by the judge in deciding protested awards are to be retained by the local association.

When protests are entertained, where the judging has been done by score card, the specimens under dispute shall be re-scored by the judge, he to act as a member of the committee of three, as provided, the re-scoring to be done in the presence of the other two members of the committee on protests.

Protests are not to be entertained except when made in writing, and the person making same shall deposit with the secretary of the local association the sum of five dollars, this money to be returned to the person making the protest if his protest be sustained; if protest be not sustained, the deposit becomes the property of the local association.

A uniform style of score card is recommended, with a view to having all associations use the same scale of points, and thus assist in making the work of the judges more accurate and uniform. The form of score card printed in this book is used and recommended by a large majority of the leading judges. It is not copyrighted, and local associations are requested to use it, each card to bear these words: "OFFICIAL SCORE CARD OF THE AMERICAN POULTRY ASSOCIATION."

Special consideration is to be given the matter of STANDARD weights and proper size at score-card shows and at comparison shows, respectively. Each specimen at all score-card shows shall be correctly weighed, regardless of circumstances. The practice of allowing a few ounces is expressly forbidden, inasmuch as it works great injustice. It has a harmful influence on the judge, the exhibitor, the local association and the industry at large, and many times deprives prudent and worthy fanciers of prizes rightfully due them.

Any exhibitor found guilty of faking, or of showing borrowed birds, shall be debarred from competition and shall forfeit any prize or prizes that may have been awarded him.

It is respectfully recommended that local associations, specialty clubs, and other organizations, advertise in their premium lists and otherwise that their exhibitions will be conducted under the rules of The American Poultry Association, and that the instructions to judges, general disqualifications, and other provisions and requirements of THE AMERICAN STANDARD OF PERFECTION shall govern.

Fraternally,

THE AMERICAN POULTRY ASSOCIATION.

CONTENTS.

G

H

Figure 1.

NOMENCLATURE

DIAGRAM OF MALE.

1 Head.	10 Hackle.	19 Primaries, Flights.	28 Body Feathers.
2 Beak.	11 Front of Hackle.	20 Primary-coverts.	29 Fluff.
3 Nostril.	12 Breast.	21 Back.	30 Thighs.
4 Comb.	13 Cape.	22 Saddle.	31-31 Hocks.
5 Face.	14 Shoulder.	23 Saddle Feathers.	32-32 Shanks.
6 Eye.	15 Wing-bow.	24 Sickles.	33-33 Spurs.
7 Wattle.	16 Wing-front.	25 Smaller Sickles.	34-34 Feet.
8 Ear.	17 Wing-coverts, Wing-bar.	26 Tail-coverts.	35-35-35 Toes.
9 Ear-Lobe.	18 Secondaries, Wing-bay.	27-27 Main Tail Feathers.	36-36 Toe Nails.

18

Figure 2.

NOMENCLATURE

DIAGRAM OF FEMALE

1 Head.	10 Neck.	19 Primaries. Flights.	28 Fluff.
2 Beak.	11 Front of Neck.	20 Primary-coverts.	29 Thighs.
3 Nostril.	12 Breast.	21 Back.	30-30 Hocks.
4 Comb.	13 Cape.	22 Sweep of Back.	31-31 Shanks.
5 Face.	14 Shoulder.	23 Cushion.	32 Spur.
6 Eye.	15 Wing-bow.	24-24 Main Tail Feathers.	33-33 Feet.
7 Wattle.	16 Wing-front.	25-25 Tail-coverts.	34-34-34 Toes.
8 Ear.	17 Wing-coverts.	26-26 Tail-coverts.	35-35 Toe Nails.
9 Ear-Lobe.	18 Secondaries, Wing-bay.	27 Body Feathers.	

19

GLOSSARY OF TECHNICAL TERMS.

BARRING: Bars or stripes extending across a feather at right angles to its length, or nearly so. (See illustrations, figures 3, and 4.)

BEARD: In chickens, a group of feathers pendant from the throat, as in Houdans and Polish. (See illustration, figure 10.)

In turkeys, a tuft of coarse, bristly hairs, four to six inches long, projecting from the upper part of breast of mature males. (See pages 360 and 366.)

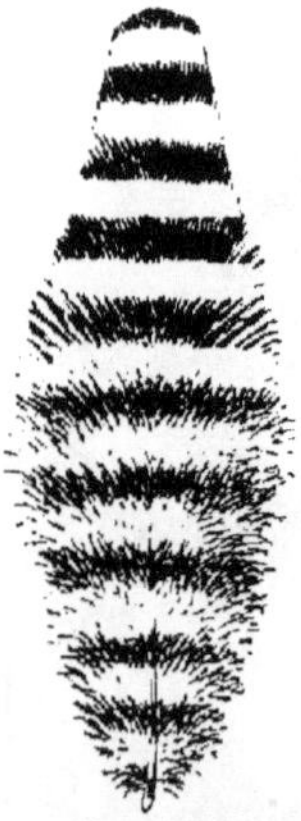

BEAN: A hard, bean-shaped protuberance growing at the tip of the upper mandible of a water fowl. (See illustration, figure 12.)

BEAK: The projecting mouth parts of chickens and turkeys, consisting of upper and lower mandible. (See illustrations 1 and 2.)

BILL: The projecting mouth parts of water fowl, consisting of upper and lower mandible. (See illustration, figure 12.)

BLADE: The rear part of a single comb, back of the last well defined point, usually extending beyond the crown of the head, smooth and free from serrations. (See illustration, figure 6.)

BRASSINESS: Having the color of brass, yellowish. A serious defect in all breeds, except Games and Game Bantams.

BREED: A race of fowls, the members of which maintain distinctive shape characteristics that they possess in common. Breed is a broader term than variety. Breed includes varieties, as, for example, the Barred, White and Buff varieties of the Plymouth Rock breed.

BREEDER: A broad, general term that designates

Figure 3.
Barred Feather
Ideal. (Female).

Figure 4.
Barred Feather. Ideal (Male).

the poultry raiser who produces fowls for any special purpose with the object of improving their value or in conformity
with an agreed standard of excellence.

CAPE: The short feathers on the back underneath the hackle,
collectively shaped like a cape.

CARRIAGE: The attitude, bearing or style of a bird.

CARUNCLES: Small, fleshy protuberances,—as on
the head of a turkey.

CARUNCULATED: Having caruncles.

CAVERNOUS: Applied to the hollow protruding nostrils of the crested breeds.

CHICKS: The young of the domestic hen, properly
applied until the sex can be distinguished; sometimes used to designate specimens less than a
year old.

CHICKENS: Specifically, the young of the domestic
hen prior to the development of adult plumage;
used as a general term to designate all domestic
fowls, except turkeys, ducks and geese.

COCK: A male fowl one year old and over.

COCKEREL: A male fowl less than one year old.

COMB: The fleshy protuberance growing on top of
a fowl's head. The Standard varieties of combs
are: Single, rose, pea, V-shaped and strawberry, all others
being modifications of these. (See illustrations, figures 6, 7,
8, 9, 10 and 11.)

CONDITION: The state of a fowl in regard to health, cleanliness
and order of plumage.

COVERTS: See tail, flight and wing coverts. (See illustrations,
figures 1 and 2.)

CREAMINESS: Having the color of cream; light yellow.

CREST: A crown or tuft of feathers on the head of a fowl. (See
illustration, figure 10.)

CROP: The receptacle in which a fowl's food is accumulated before it passes to the gizzard.

CUSHION: The mass of feathers at the rear of back of a fowl,
partly covering the tail; most pronounced in Cochin females.
(See Cochin illustrations.)

DEWLAP: A pendulous skin development under throat.

DISQUALIFICATION: A deformity or serious defect that renders
a fowl unworthy to win a prize.

DISQUALIFIED: Applied to a fowl that is unworthy to win a prize.

Figure 5.
Penciling Par-
allel   Form.
(Ideal).

21

Figure 6.
One Type of Single Comb (Ideal).
1, Base; 2, 2, 2, 2, 2, Points; 3, Blade.
For Other Types See Illustrations of
Plymouth Rocks, Minorcas, Javas,
Orpingtons, etc.

Figure 7.
One Type of Rose Comb (Ideal).
1, Base; 2, Rounded Points; 3, Spike.
See Figures 1 and 2 for Ideal Wyandotte Combs.

Figure 8.
Pea Comb, Profile (Ideal).

Figure 9.
Pea Comb, Quarter View (Ideal).

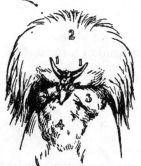

Figure 10.
Sultan's Head, Male (Ideal). 1-1,
V-shaped Comb; 2, Crest; 3, 3, Muffs;
4, Beard.

Figure 11.
Strawberry Comb (Ideal).

22

DOWN: The first hairy covering of chicks; also the tufts of hair-like growth that are sometimes found on the shanks, toes, feet or webs of feet of fowls.

(*Note:* If the quill and web are discernible to the eye, it is a "feather.")

DRAKE: A male of the duck family.

DUBBING: Cutting off the comb, wattles and ear-lobes, so as to leave the head smooth.

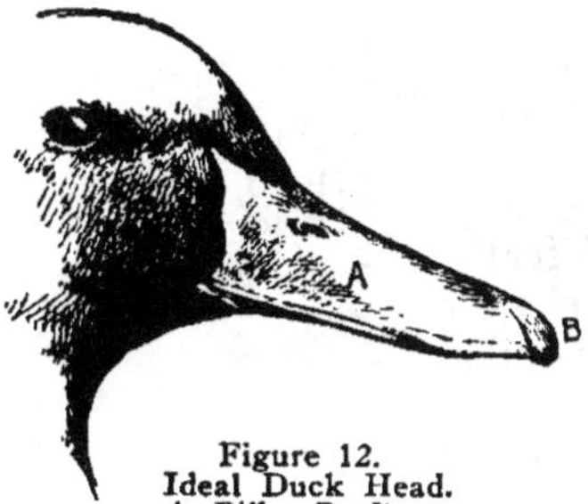

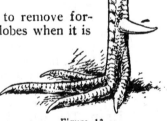

Figure 12.
Ideal Duck Head.
A, Bill. B, Bean.

DUCK: A female of the duck family, as distinguished from the drake or male.

DUCKLING: The young of the duck family in the downy stage of plumage.

DUCK-FOOTED: The hind toe carried forward. (See illustration, figure 13.)

EAR-LOBES: The folds of bare skin just below the ears. Ear-lobes of different breeds vary in color; being red, white, purple, cream, etc.; they also vary greatly in size. (See illustrations, figures 1, 2, 6, 7, 8, 9, 10 and 11.)

EXCRESCENCES: A disfiguring, abnormal or superfluous outgrowth.

FACE: The bare skin on the head of a fowl around and below the eyes. (See illustrations, figures 1 and 2.)

FAKING: Removing, or attempting to remove foreign color from the face or ear-lobes when it is a disqualification; removing one or more side sprigs; trimming a comb in any manner, except the dubbing of Games; artificial coloring or bleaching of any feather or feathers; splicing feathers; injuring the plumage of any

Figure 13.
Duck Foot (A Defect).

fowl entered by another exhibitor; plugging up holes in legs of smooth-legged varieties where feathers or stubs disqualify; staining of legs; in fact, any self-evident attempt on the part of an exhibitor to deceive the judge and thus obtain an unfair advantage in competition.

FANCIER: A breeder of poultry who seeks to produce chickens, turkeys, ducks or geese in conformity with an ideal or prescribed standard of excellence.

FAWN: The color of a young deer.

FEATHER: A growth formed of a discernible quill and a vane (called "web") upon each side of it. (See illustration, figure 14.)
(*Note*: When quill is not discernible to the eye, it is "down.")

FLIGHTS: The primary feathers of the wing, used in flying, but out of sight, or nearly so, when wing is folded. (See illustration, figure 38.)

FLIGHT-COVERTS: The short, moderate stiff feathers located at the base of the wing primaries, or flight feathers, and partly covering their quills. (See illustration, figure 38.)

FLUFF: The soft feathers about thighs and posterior part of fowl; also the soft downy part of a feather. (See illustrations, figures 1, 2 and 14.)

Figure 14.
Sections of a Feather.

FOREIGN COLOR: Any color on a feather that differs from the color prescribed for such feather as a part of the plumage of a Standard-bred fowl.

FROSTING: A marginal edging or tracing of color on feather of laced, spangled and penciled varieties. (See illustration, figure 15.)

GIPSY COLOR: Dark purple, approaching black.

HACKLE: The neck plumage of males, formed of the hackle feathers. (See illustration, figure 1.)

HACKLE FEATHERS: The long, narrow feathers growing on the neck of male. (See illustrations, figures 1, 4 and 17.)

HANGERS: A term sometimes applied to the smaller sickles and tail coverts of male. (See illustration, figure 1.)

HEAD: That part of a fowl composed of skull and face, to which the comb, crest, beak, wattles and ear-lobes are attached.

HEN-FEATHERED: A male bird that resembles a hen, owing to the absence of sickles, pointed hackle feathers, etc., is said to be "hen-feathered."

HOCK: See "Knee-Joint." (Also illustration, figure 1.)

Figure 15.
Frosting (A Defect).

IRIDESCENT: A prismatic play of color.

KNEE-JOINT: The joint between the thigh and shank.

KNOCK-KNEED: A deformity in which the legs come too near together at the knee-joints and are bent outward, laterally, below the knees.

LACED—LACING: A feather edged or bordered with a band of color, different from the ground color of feather. (See illustration, figure 16.)

LEAF COMB: A combination of two small single combs, having serrated leaf-like edges; the original Houdan comb, now replaced in America by the V-shaped comb. (See illustration, figure 10.)

LEG: Includes thigh and shank. (See illustrations, figures 1 and 2.)

LEG-FEATHERS: Feathers growing on the outer side of the shank, as in Asiatics. (See illustration, figure 21.)

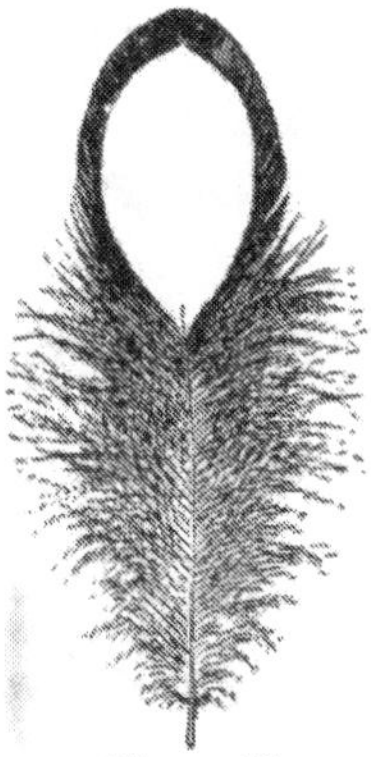

Figure 16.
Laced Feather
(Ideal).

LOPPED COMB: A comb falling over to one side. To disqualify (see "General Disqualifications") a single comb, some portion must fall below the horizontal plane where the comb begins to lop; a rose or pea comb, it must lop over far enough to come in contact with one side of the head or obstruct the sight. (See illustrations, figures 19 and 20.)

LUSTER: The special brightness of plumage that gives brilliancy to the surface color of the fowl or section.

MEALY: Having the appearance of being sprinkled with

Figure 17.
Striped Neck (Hackle)
Feather, Male (Ideal).

Figure 18.
Striped Neck Feather,
Female (Ideal).

25

Figure 19.
Lopped Single Comb (A
Disqualification).

Figure 20.
Lopped Rose Comb (A
Disqualification).

Figure 21.
Cochin Leg and Toe Feathering. A, Up-
per Thigh: B. B, Lower Thigh; C, C,
Shank; D, D, Toe.

meal. Applied to buff or red varieties where the ground color is stippled with a lighter color. (See "Stipple"; also illustration, figure 22.)

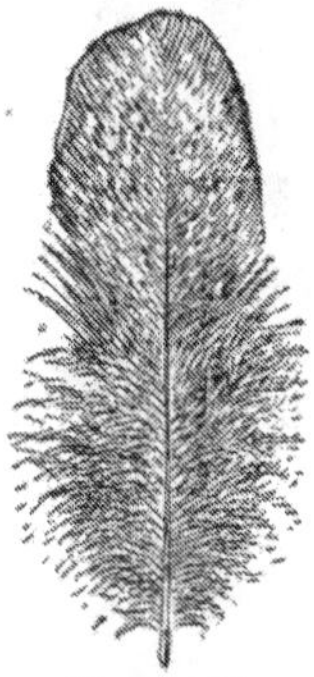

MOSSY: Irregular dark penciling appearing in feathers and destroying the desirable contrast of color. (See illustration, figure 23.)

MOTTLED: Marked on the surface with spots of different colors or shades of color.

MUFFS: The cluster of feathers covering the sides of the face below the eyes, extending from the beard to the ear-lobes and found only on bearded varieties. (See illustration, figure 10.)

NOSTRILS: Openings beginning at base of beak and extending into the head.

OBTUSE ANGLE: An angle greater than a right angle, i. e., one containing more than ninety degrees.

Figure 22.
Mealy (Defective) Feather.

PARTI-COLORED: Feathers or fowls having two or more colors.

PEA COMB: A triple comb of medium length, resembling three straight single combs placed parallel with one another and joined at base and rear, each having short but distinctly divided serrations, the serrations of the two outer rows being lower and smaller than those of the middle row, and those of each row being somewhat larger and thicker midway of the comb than at front and rear. (See illustrations 8 and 9.)

PEN: (Exhibition.) A male and four females of the same variety.

PENCILING: Small markings or stripes on a feather. They may run straight across, as in the Penciled Hamburgs, in which case they frequently are called "Bars." or may follow the outline of the feather, taking a crescentic form, as in the Dark Brahmas, Partridge Cochins, etc. (See illustrations, figures 5 and 24.)

Figure 23.
Mossy (Defective) Feather.

PEPPERED—PEPPERING: Sprinkled with gray or black. (See "Mealy.")

PINION FEATHERS: The feathers attached to the joint of the wing that is most remote from the body.

PLUMAGE: The feathers of a fowl.

POULT: The young of the domestic turkey, properly applied until sex can be distinguished, when they are called cockerels and pullets.

POULTRY: Domestic fowls reared for exhibition, for the table, or for their eggs or feathers.

PRIMARIES: (See "Flights" and illustrations, figures 1, 2 and 38.)

PROFILE: A direct side view of a fowl. Applied to live specimens and to illustrations.

PULLET: A female fowl less than a year old.

QUILL: The hollow, horny, basal part or stem of a feather. (See "Shaft"; also illustration, figure 14.)

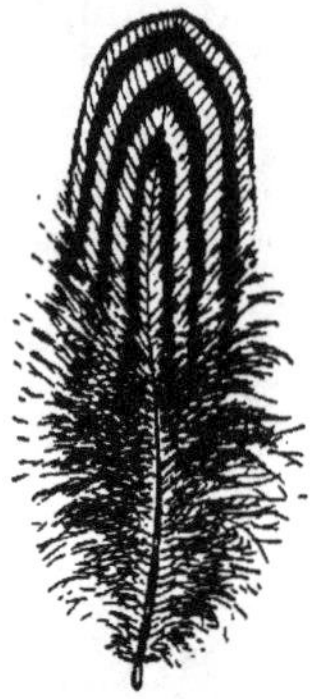

Figure 24.
Penciling. Crescentic Form (Ideal).

ROSE COMB: A low, solid comb, the upper surface free from hollow center and covered with small rounded points. This comb terminates in a well-developed spike which may turn upward, as on Hamburgs; be nearly level, as on Rose Comb Leghorns; or turn downward, as on Wyandottes. (See illustrations, figures 1, 2, 7 and 20.)

RUMP: The rear portion of the back of a duck or other fowl.

SADDLE: The rear part of the back of a male bird, extending to the tail and covered by the saddle feathers. (See illustration, figure 1.)

SADDLE FEATHERS: The feathers growing out of the saddle. (See illustration, figure 1.)

SADDLE HACKLE: The long, narrow, pointed feathers growing from a male bird's saddle and drooping at the sides.

SCALY LEGS: Incrustations or deposits upon and beneath the scales of a fowl's legs.

SCOOP BILL: A basin like cavity in the center of bill of a water fowl. (See illustration, figure 25.)

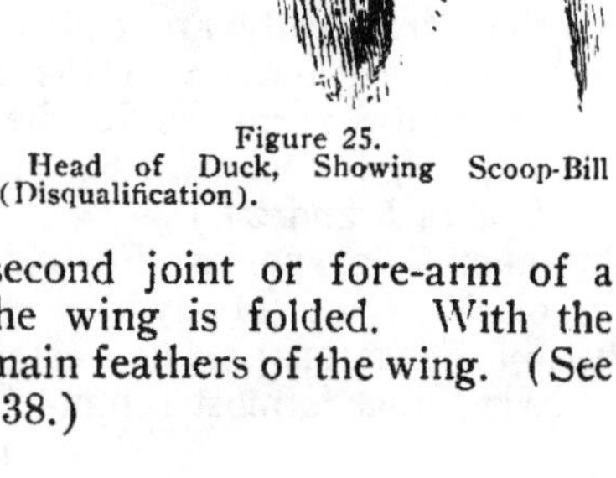

Figure 25.
Head of Duck, Showing Scoop-Bill (Disqualification).

SECONDARIES: The long quill feathers that grow on the second joint or fore-arm of a fowl's wing, visible when the wing is folded. With the primaries, they constitute the main feathers of the wing. (See illustrations, figures 1, 2, and 38.)

28

SMALLER SICKLES: See "Sickles."

SOLID COLOR—SELF COLOR: A uniform color, unmixed with any other.

SERRATED: Notched along the edge like a saw.

SERRATION: A V-shaped notch between the points of a single comb.

SHAFT: The stem of a feather, especially the part filled with pith, which bears the barbs. (See illustrations, figures 14 and 26.)

SHAFTING: The shaft of the plume portion of a feather, being lighter or darker in color than the web of the feather. (See illustrations, figures 14 and 26.)

SHANK: The portion of a fowl's leg below the hock, exclusive of the foot and toes. (See illustrations, figures 1 and 2.)

SICKLES: The long, curved feathers of the male bird's tail, properly applied to the top pair only, but sometimes used in referring to the prominent tail-coverts, which are also called smaller sickles. (See illustrations, figures 1 and 2.)

SIDE SPRIG: A well-defined, pointed growth on the side of a single comb. (See illustration, figure 27.)

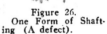

Figure 26.
One Form of Shafting (A defect).

SINGLE COMB: A comb consisting of a single, thin, fleshy, serrated formation, rising from the beak and extending backward over the crown of the head, and in males, beyond the head. (See illustration, figure 6.)

SLIPPED WING: A wing of a fowl not closely folded and held up in proper position; a defect resulting from injury or from weakness of muscles of wing. (See illustration, figure 29.)

Figure 27.
One Form of Side Sprigs (A Disqualification).

SPANGLE: A clearly defined marking of a distinctive color, located at the end of a spangled feather. (See illustration, figure 30.)

SPANGLED: Plumage made up of spangled feathers.

SPLASHED FEATHER: A feather with colors scattered and irregularly intermixed. (See illustration, figure 31.)

29

Figure 28.
Split Comb. Showing the Tendency of the Blade to Divide Perpendicularly (Disqualification).

SPLIT COMB: A single comb which is divided perpendicularly and the two parts overlap. (See illustration, figure 28.)

SPUR: A horn-like protuberance, growing from the inner side of the shank of a fowl; may be knob-like or pointed, according to the age of the fowl and the sex. (See illustration, figure 1.)

SQUIRREL TAIL: A fowl's tail, any portion of which projects forward, beyond a perpendicular line drawn from the juncture of tail and back. (See illustration, figure 32.)

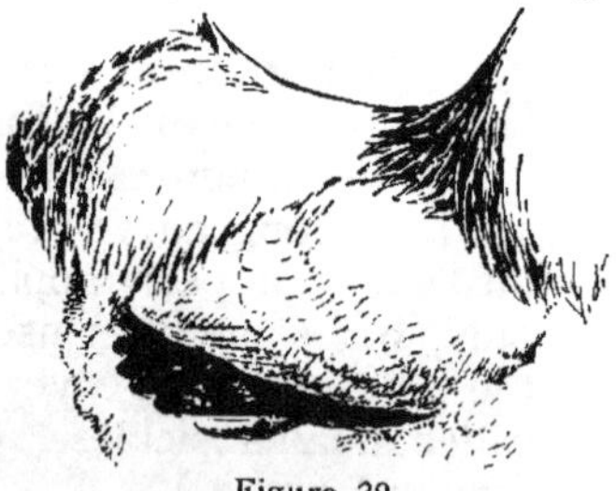

STATION: Ideal pose, embodying Standard style, notably height and reach, as applied to Games.

STERN: The lower or under part of the posterior section of a fowl.

STIPPLE: Verb, to execute on stipple, i. e., draw, paint or engrave by means of dots instead of lines. Noun, the effect obtained in color work by the use of dots instead of strokes or lines. . (See illustration, figure 33.)

Figure 29.
Slipped Wing and Twisted Feather (Defects).

STRAIN: A family of any variety of fowls bred in line by descent by one breeder, or successor, during a number of years, that has acquired individual characteristics which distinguish it more or less from specimens of other strains of the same variety.

STRAWBERRY COMB: Approaching in shape the outline and surface of a strawberry. (See illustration, figure 11.)

STRIPE: A line or band of color, regular or irregular in form, that differs from the body color.

STRIPED FEATHER: A feather, the surface of which contains lines or bands of color, regular or irregular in form, differing from the body color. (See illustrations, figures 17 and 18.)

Figure 30.
Spangled Feather (Ideal).

STUB: A short feather, or portion of a feather, when found between or under scales of shanks or toes.

SURFACE COLOR: The color of that portion of the plumage of a fowl that is visible when the feathers are in their natural position.

SYMMETRY: Perfection of proportion; the harmony of all parts or sections of a fowl, viewed as a whole, with regard to the Standard type of the breed it represents.

TAIL-COVERTS: The curved feathers in front of and at the sides of the tail. (See illustration, figure 1.)

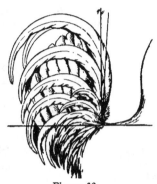

Figure 32.
Squirrel Tail. (A Disqualification except in Japanese Bantams).

TAIL FEATHERS: Main, the straight and stiff feathers of the tail that are contained inside the sickles and tail coverts; the top pair are sometimes slightly curved, but are generally straight. (See illustrations, figures 1 and 2.)

THIGHS: That part of the legs above the shanks. (See illustrations, figures 1 and 2.)

THUMB-MARKS: A disfiguring depression which sometimes appears in the sides of a single comb. (See illustration, figure 34.)

TICKING: The specks or small spots of black color on the tips of neck feathers of Rhode Island Red females: small specks of color on feathers that differ from the ground or body color.

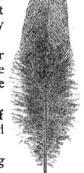

Figure 31.
Splashed (Defective) Feather.

TIPPED: A term applied to a feather ·the web end of which differs in color from the color of the body or main portion of the feather.

TOE FEATHERING: The feathers on the toes of fowls required to have feathered shanks and toes. (See illustration, figure 21.)

TOP-KNOT: A word wrongly used as meaning crest; no longer employed. (See illustration, figure 10.)

TRIO: One male and two females.

Figure 33.
Stippled Feather (Ideal).

TWISTED COMB: An irregularly shaped comb, falling or curving from side to side, being distorted from the normal perpendicular position. (See illustration, figure 35.)

TWISTED FEATHER: Feather with quill or shaft twisted. (See illustration, figure 29.)

TYPICAL: Expressing a characteristic in color or form, representative of a breed or variety; for example, typical shape, meaning the form peculiar to a breed.

Figure 34.
Thumb Marks in Comb, Rear Turning to One Side. (Defects).

Figure 35.
One Form of Twisted Single Comb (Defect).

UNDER-COLOR: The color of the downy portion of the plumage, not visible when the plumage of the fowl is in natural position. (See illustration, figure 14.)

VARIETY: A sub-division of a breed (see definition of "Breed"), used to distinguish fowls having the Standard shape of the breed to which they belong, but differing in color of plumage, shape of comb, etc., from other groups of the same breed. The general difference between the terms "Breed" and "Variety" is well brought out in the statement popular among breeders and fanciers: "Shape makes the breed; color, the variety."

V-SHAPED COMB: A comb formed of two well defined, horn-shaped sections. (See illustration, figure 10.)

VULTURE-HOCK (Vulture-Feathered): The stiff quill feathers growing on the thighs, extending backward, straight beyond the knee-joint, or "hock"; to disqualify, they must be without a sufficient quantity of fluffy feathers to relieve the stiff appearance and fill up the sharp angles, viewed in profile. (See illustration, figure 36.)

Figure 36.
Vulture-Like Hocks (As shown, a Disqualification).

32

WATTLE: The pendant growths at the sides and base of beak.

WEB: Web of Feather: The flat portion of a feather, made up of a series of barbs on either side of the shaft. (See illustration, figure 14.) Web of Feet: The flat skin between the toes. Web of Wings: The triangular skin attaching the wing to the body, visible when wing is extended.

WING-BAR: The stripe or bar of color extending across the middle of the wing, formed by the color or marking of the wing-coverts. (See illustration, figure 1.)

WING-BAY: The triangular section of the wing, below the wing-bar, formed by the exposed portion of the secondaries when the wing is folded. (See illustrations, figures 1 and 2.)

WING-BOW: The upper or shoulder part of the wing. (See illustrations, figures 1 and 2.)

WING-COVERTS: The small, close feathers clothing the bend of the wing and covering the roots of the secondary quills. (See illustrations, figures 1, 2 and 38.)

WING-FRONTS: The front edge of the wing at the shoulder. This section of the wing is sometimes called "wing-butts." The term wing-fronts is recommended, thus avoiding confusion. (See illustrations, figures 1 and 2.)

WING-POINTS: The ends of the primaries, erroneously called "wing-butts." (See illustrations, figures 1, 2 and 38.)

WRY TAIL: Tail of a fowl turned to one side, permanently so. (See illustration, figure 37.)

Figure 37.
Showing Wry-Tail (Disqualification).

....

(Names of Association, here)

....

(Date; month, days and year show is held, here)

OFFICIAL SCORE CARD OF THE AMERICAN POULTRY ASSOCIATION

Exhibitor ....

Variety .... *Sex* ....

Entry No....................*Band No*....................*Weight*....................

	Shape	Color	Remarks
Symmetry			
Weight or Size......			
Condition			
Comb			
Head			
Beak			
Eyes			
Wattles and Ear-Lobes			
Neck			
Wings			
Back			
Tail			
Breast			
Body and Fluff..............			
Legs and Toes..............			
Crest and Beard..........			
†*Shortness of Feather*..			
Total Cuts................		Score..............	

...., *Judge*

...., *Secretary*

*Applies to Crested Breeds.   †Applies to Games and Game Bantams.

This card to be printed on card board 3½x6½ inches and is here printed the exact size and the above is a fac-simile.

Score cards may be obtained from the Secretary of the American Poultry Association.

34

INSTRUCTIONS TO JUDGES.

MERIT: The merit of specimens shall be determined by a careful examination of all sections in the "Scale of Points," beginning with symmetry and continuing through the list, deducting from the full value of each section of a perfect specimen, for such defects as are found in the specimen. Judges must familiarize themselves with the scale of points of each breed they are to pass upon, to intelligently award prizes. And it must be understood that no more and no less value can be placed on any section than is provided for in the "Scale of Points." And it shall be further understood that this system must be applied whether judged by score card or comparison. The minimum cut for any section shall be one-fourth of one point.

WEIGHT: (a) All specimens shall be judged according to their Standard weights, provided, however, that the disqualifying weight for chicks and poults shall not apply until December first of each year. (b) In all breeds of fowls having weight clauses, except Bantams, deduct two points per pound for amount lacking from standard weights, and in that proportion for any fractional part of a pound, using one-fourth pound as a minimum, the specimen to have the benefit of any fraction less than one-fourth pound. (c) In all varieties of Bantams, deduct one-half point per ounce for any excess over standard weights. (d) In all varieties of turkeys and geese, and in all varieties of ducks, except those prized for their smallness, having weight clauses, deduct three points per pound for amount lacking from standard weights, and in that proportion for any fractional part of a pound, using one-fourth pound as the minimum, the specimen to have the benefit of any fraction less than one-fourth pound. (e) In all varieties of fowls, except Bantams, also in all varieties of turkeys and geese, and all varieties of ducks except those prized for their smallness, when adult specimens are equal in score and are above or below standard weight, the one nearest weight shall be awarded the prize, except when one specimen is cut for weight, and the others are not, the specimen that is standard weight or above shall be awarded the prize. (f) All chicks or immature specimens—except Bantams and those varieties prized for their smallness,—having an equal score, when cut for lack of weight, the one of less weight shall be awarded the prize; but when each of such specimens are of standard weight, or over, the one nearest weight shall be awarded the prize. (*g*) In all varieties of Bantams, and those varieties of ducks prized for their smallness, other things being equal, the smallest specimen shall win. (CAUTION—The weight clause must not be understood to mean that a small, but over-fat specimen is within the spirit of the meaning of the Standard; the size must be proportionate to the weight, preserving the ideal shape and type of the Standard specimen.)

RE-WEIGHING: The j u d g e may, at his option, demand the re-weighing of the specimens in competition, in all classes where Standard weights apply.

SIZE: Size shall be determined by comparison of the specimens. In all varieties (except Call and East India Ducks), having a section termed "Size," and not being subject to weight clauses, the largest specimen, other things being equal, shall win; in Call and East India Ducks this rule is reversed and the smallest specimens shall win.

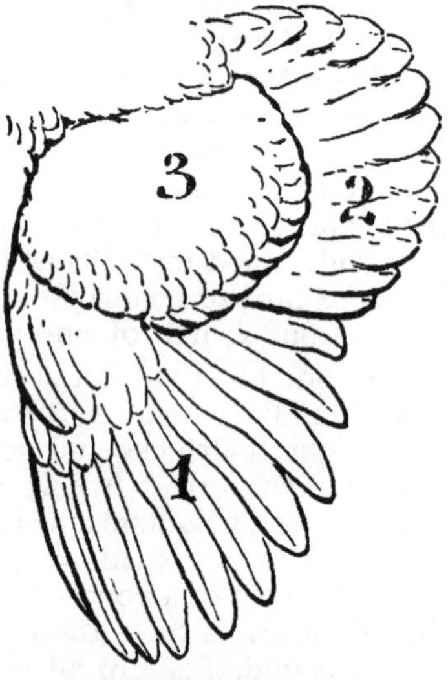

Figure 38.
Showing Divisions of Wing.
1 Flights or Primaries.   2 Secondaries.
3 Fronts, wing-bows and bar.

WING DIVISION: In discounting the color of wings, the section should be divided into three separate parts, allowing two points for fronts, wing-bow and bar; two for primaries and primary-coverts; two for secondaries; and no greater value can be placed on any one of these parts. (See illustration, figure 38.)

SCORES ENTITLING SPECIMENS TO PRIZES: To receive a first prize the specimen must score 90 points or more, except

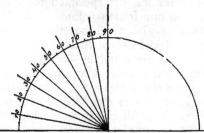

Figure 39.
Diagram Showing Degrees from Horizontal.

36

cocks of all parti-colored varieties, which may be awarded first prize, provided they score 88 points or more. For each receding prize drop one point. A pen to win first prize must score 180 points or more, unless it contains a cock of a parti-colored variety, in which case 178 points or more may win first prize; but first prize shall not be given on a pen if the male in pen scores less than 88 points. No prize shall be awarded an exhibition pen if any specimen in the pen scores less than 85 points.

Judges shall not be required to score turkeys or water fowls.

SWEEPSTAKE PRIZES: In competition for sweepstake prizes, when solid-colored specimens compete with parti-colored specimens, white specimens shall be handicapped two points each, black specimens one and one-half points each, and buff specimens one point each; after such reduction, the specimen having the highest score, or the specimens having the highest average or combined score shall be awarded the prize.

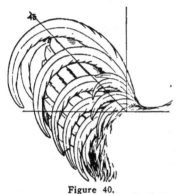

Figure 40.
Tail Carried at Angle of 45 Degrees.

Bantams, water fowl or turkeys are not eligible to compete for sweepstake prizes.

OLD AND YOUNG SPECIMENS: All other points being equal, where prizes are offered on old and young specimens competing together, the former shall be awarded the prizes.

FAKING: Faking of any description shall debar from competition specimens so treated. (See Glossary for what is meant by "Faking.")

CREAMINESS AND BRASSINESS: In White varieties, except where the color of plumage is specified as creamy-white, the presence of brassiness on surface, or creaminess of quills or under-color, is a serious defect and is to be discounted accordingly.

Bleaching by means of chemicals is such a harmful practice that where it is proven by other evidence than the condition of the specimen or specimens, such bleached specimen or specimens shall be considered faked and disqualified.

SCORE OF EXHIBITION PEN: To ascertain the score of an exhibition pen, add the scores of the females together and divide the sum by the number of females in the pen; to the quotient thus obtained add the score of the male, and this sum shall be the score of the exhibition pen.

DATED SCORE CARDS: All score cards made out by judges applying the Standard are to be dated with ink, indelible pencil or stamp on the date the specimens are judged.

DEFECTIVE SCORE CARD: It shall be considered irregular for a judge to sign a score card unless the weight of all breeds and varieties having Standard weights is considered, regardless of the season.

PRIVATE SCORING: Private scoring of specimens is not advisable and members of this AS-

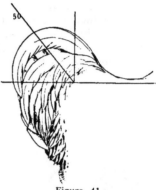

Figure 41.
Tail Carried at Angle of 50 Degrees.

SOCIATION are directed not to lend their support to the practice as a selling method. Judges are ordered to weigh each specimen and apply the proper cut and to make proper cuts for the condition of the specimen at the time the fowl is scored.

TIES: In case of ties between two or more specimens that cannot be broken by any of the previous rules, the specimen receiving the smallest total sum of cuts for shape shall be awarded the prize. In case of ties on exhibition pens, when the tying pens contain either all old or all young specimens, the adult pen shall win; when the tying pens are both adult or both young, the pen containing the highest scoring male shall win; when the tying pens contain females of mixed ages, the pen containing the highest scoring male shall win; when one of the pens contains all hens or all pullets, while the other contains females of mixed ages, the pen having all the females either adult or young shall win; when the tie cannot be broken by any of the above rules, the pen containing the lowest total of shape cuts in the five main shape sections shall win.

IN APPLYING THE COMPARISON SYSTEM.

TYPICAL SHAPE: In awarding prizes by comparison, judges must consider carefully each and every section of the specimen, according to the Scale of Points and not allow color alone, or any one or two sections to influence their decisions. The vital importance of typical shape is to be borne constantly in mind, at the same time giving due consideration to color in all sections, including under-color.

HANDLING: All specimens in competition must be handled and examined by the judge, except those that show decided inferiority as seen in coops.

DISQUALIFYING WEIGHTS: Specimens falling below disqualifying weights after December first of each year must be debarred from competition, except Bantams, which, when exceeding disqualifying weights, shall suffer a like penalty.

STANDARD SIZE: In determining size, the judge shall decide by comparing the specimens in competition with due regard to weight in all breeds and varieties, where weight is required by the Standard. When a bird fails to attain, or in case it exceeds, the size proportionate with the type or shape, it must be discounted quite severely.

BANTAM TYPE: While smallness of size is desirable in all Bantams, no specimen shall be entitled to win over a larger bird simply because of its small size; it must conform to the type demanded for the breed it represents.

GAME TYPE: In judging Games, Game Bantams, and Sumatras, it is imperative that shape be considered of greatest importance. Specimens lacking in this essential breed characteristic shall not be awarded first honors, even if there be no competition.

COLOR DEFECTS: A few, very small, grayish specks in white fowls shall not debar a specimen, that is otherwise superior in color, from winning over one less typical in shape and sound in color; provided, however, that the gray specks do not appear prominently in the primary, secondary or main tail feathers.

SCALY LEGS: A fowl whose legs and toes are so deformed by what is called "Scaly Legs," as to hide or appear to have destroyed the color, shall not be awarded a first prize.

NOTE: Under the comparison system, judges must deduct the full valuation of the cuts in all sections, where a specified cut is made under the heading of "Cutting for Defects."

GENERAL DISQUALIFICATIONS.

NOTE: If, in applying the Standard of Perfection, judges find any of the defects described below, they shall disqualify the specimen and state on the proper card or blank the nature of the disqualification:

Specimens unworthy of a score or lacking in breed characteristics.

39

In the Asiatic breeds, except Langshans, and in Cochin Bantams and Booted White Bantams, shanks not feathered down outer sides, outer toes not feathered to the last joint. In Cochin and Cochin Bantams bare middle toes.

In Langshans, shanks not feathered down the outer sides; feathers not growing beyond the middle joint of the outer toe.

In Silkies and Sultans, shanks not feathered down the outer sides.

In all breeds required to have unfeathered shanks, any feather or feathers, stubs or down on shanks, feet or toes; or unmistakable indications of feathers, stubs or down having been plucked from same.

Plucked hocks.

Webb feet in any breed of chickens.

In four-toed breeds, more or less than four toes on either foot. In five-toed breeds, more or less than five toes on either foot.

Legs and toes of color foreign to the breed.

A wing showing clipped flights or secondaries, or both, except in wild geese or ducks, shall disqualify the specimen and debar it from competition.

Deformed beaks.

Decidedly wry tails.

Crooked backs.

In all Ducks and Geese, twisted wings, crooked backs, decidedly wry tails.

Lopped Combs, except in Mediterranean, Continental and Dorking females. Rose Combs falling to one side, or so large as to obstruct sight. Combs foreign to the breed. Split combs. (See illustrations, figures 19, 20 and 28.)

The comb on a specimen which merely turns over a trifle from the natural upright position is not to disqualify.

Side sprig or sprigs on all single comb varieties. (See illustration, figure 27.)

Absence of spike in all rose comb varieties, except Silkies, Malays and Malay Bantams.

Entire absence of main tail feathers.

Decidedly squirrel tail in all breeds, except Japanese Bantams.

Positive enamel white in the face of Mediterranean cockerels and pullets except White Faced Black Spanish.

In varieties where positive enamel white in ear-lobes is a

disqualification, judges shall disqualify for unmistakable evidence of an attempt to remove the defect.

Absence of crest or beard in any variety described as Crested or Bearded, or any appearance of crest or beard in any variety where not called for.

Absence of knob in young, or absence of knob or dewlap in adult specimens of African Geese.

Absence of knob in Chinese Geese.

In any breed having weight clauses, except Turkeys and Bantams, a specimen falling more than two pounds below Standard weight.

In all varieties of Turkeys, specimens falling more than six pounds below Standard weight.

In all varieties of Bantams, specimens more than four ounces above Standard weight.

Black in the bean or bill of Pekin or Aylesbury drakes. (See illustration, figure 12.)

Faking in any manner shall disqualify the specimen.

Under all disqualifying clauses, the specimen shall have the benefit of the doubt.

NOTE: In all varieties of fowls, red pigment on sides or back of shanks not to be considered a defect.

CUTTING FOR DEFECTS.

These cuts should not be confused with nor take precedence over the valuation given each section in the Scale of Points of all varieties.

Judges, in applying the score card, are to discount for the more common defects, as follows:

	Points
Frosted Combs	½*
Too many or too few points on single combs, each...	½
Thumb mark on comb, not less than	1
Rear of comb turning around..................	½ to 1
Coarse texture of comb	½ to 1
Roughness, irregularity, hollow center, over-size and ill-shape in comb of rose-comb varieties, each defect	½ to 2
Gray or White in any except disqualifying sections of plumage of all Partridge varieties and Brown Leghorns, cut	½†
Lack of luster on surface in red and black varieties, in each section calling for luster	½

*To shape limit. †To color limit.

More than one spike on rear of rose-comb, each.... 1

Coarse texture of wattles........................ ½ to 1

For missing feather or part of feather in primaries or secondaries, where foreign color disqualifies.... 1 to 3

Where feather is broken, but not detached, in primaries or secondaries, where foreign color disqualifies ½

For broken or missing feather or feathers in primaries or secondaries of buff or parti-colored varieties, where foreign color does not disqualify........ ½ to 1

Absence of sickles, where foreign color disqualifies, for each sickle............................... 1 to 1½

Absence of sickles, where foreign color does not disqualify, for each sickle 1

Absence of one or more main tail feathers in varieties subject to color disqualifications, each...... 1

Absence of one or more main tail feathers, when not a disqualification, each .... ½

For twisted feather or feathers, in wing or tail of any variety, excepting water fowls when this disqualifies.................................... 1 to 2

Feathered middle toes in Langshans .... ½ to 1½

Brassiness in all varieties, in each section where found 1 to 2

Creaminess of plumage or quill in white varieties, except where specified creamy white, in each section where found .... ¼ to 1½

Purple barring in plumage of any variety in each section where found............................. ½ to 2

Frosty edging in any laced section of laced or spangled varieties, in each section where found.......... ¼ to 1½

Irregular, indistinct, crescentic or too heavy lacing in laced varieties, in each section where found..... ½ to 1½

Irregular barring in Barred Plymouth Rocks, in each section where found .... ½ to 1½

Light colored shafting in buff or red varieties, in each section where found .... ½ to 1½

Gray specks in any part of plumage of white varieties, in each section where found.............. ½ to 2

Mealiness in plumage of buff or red varieties, in each section where found .... 1 to 1½

Mossy-centered feathers in laced varieties, in each section where found .... ½ to 2½

Irregular or deficient penciling in penciled varieties, in each section where found.................... ½ to 1½

Black or white in buff varieties, in each section where found, cut from one-half point to the color limit of the sections.

Slate under-color in buff and red varieties, in each section where found .... ½ to 1½

Color of eyes not as described for the different varieties .... ½ to 1½

Red eyes in Campines. Cut to color limit.

If eye is destroyed, leaving only the socket......... 1½

If eye shows permanent injury, but retains its form. ½ to 1

Ear-lobes of Wyandottes showing any positive enamel white .... ½ to 2

Any positive white in the ear-lobes of any variety of Cochins .... ½

Positive white covering one-third or more of the surface of ear-lobes of any variety of Cochins ..... 1 to 3

Red markings directly above the eyes of White-Faced Black Spanish .... ½ to 2½

For positive white in face of cocks of Mediterranean class, except in White-Faced Black Spanish .... ½ to 2½

Pinched or "Gamy" tails in Leghorn females........ ½ to 1½

If tail in any specimen shows not to exceed three-fourths development 1

If tail in any specimen shows not to exceed one-half development 2

If tail in any specimen shows not to exceed one-fourth development 3

For black in bean of white ducks (females), except White Muscovys 1

For black in bill of white ducks (females), except White Muscovys, other than black in bean..... 1 to 1½

Crooked breast bone .... ½ to 2

For each bare middle toe in Brahmas 1

Crooked toes, each .... ½ to 1

In Barred Plymouth Rocks for black feather or feathers, in each section where found .... ½ to 1½

White or Gray barring in main tail feathers of Bronze Turkeys .... ½ to 2

43

DESCRIPTION OF BREEDS.

CLASS I.
AMERICAN.

Breeds — *Varieties*

PLYMOUTH ROCKS.........................
- Barred
- White
- Buff
- Silver Penciled
- Partridge
- Columbian

WYANDOTTES.............................
- Silver
- Golden
- White
- Buff
- Black
- Partridge
- Silver Penciled
- Columbian

JAVAS..................................
- Black
- Mottled

DOMINIQUES..............................

RHODE ISLAND REDS......................
- Single Comb
- Rose Comb

BUCKEYES...............................

SCALE OF POINTS.

Symmetry	4
Weight	4
Condition	4
Comb	8
Head — Shape 2, Color 2	4
Beak — Shape 2, Color 2	4
Eyes — Shape 2, Color 2	4
Wattles and Ear-lobes — Shape 2, Color 2	4
Neck — Shape 4, Color 6	10
Wings — Shape 4, Color 6	10
Back — Shape 5, Color 5	10
Tail — Shape 5, Color 5	10
Breast — Shape 5, Color 5	10
Body and Fluff — Shape 5, Color 3	8
Legs and Toes — Shape 3, Color 3	6
	100

PLYMOUTH ROCKS

Plymouth Rocks are classified as "general purpose fowls." The pioneer variety, the Barred Plymouth Rock, then called Plymouth Rock, was first exhibited in 1869 at Worcester, Mass. They are a composite of several different blood lines, the first and most prominent of which were the Black Cochin and Dominique.

In size the Plymouth Rock is intermediate between the Asiatic and Mediterranean breeds, the most typical and useful specimens are those which are nearest to Standard weights.

The six varieties are identical, except in color. The color of the Barred variety is exceedingly difficult to describe; in fact, the true and exact shades can be learned only by observation; the colors should be modified black and white in all sections, each feather crossed by regular, narrow, parallel, sharply defined, dark bars that stop short of positive black; the overlapping of the feathers producing a bluish tinge when viewed under certain light reflections.

The White variety—plumage pure white, as the name indicates, should be free from creaminess and brassiness. The combination of pure white plumage, with bright red comb, face, wattles and ear-lobes, and yellow legs and beak is both desirable and obtainable.

The color of plumage of the Buff variety should be a rich golden-buff, free from shafting or mealy appearance, while extremes of light and dark shades should be avoided, and a harmonious blending of buff in all sections is most desired.

The contrast of black with white in males and with steel-gray in females will attract many to the Silver Penciled variety. The exquisite penciling with the rich plumage and mahogany surface of the Partridge female and the brilliant red and greenish-black plumage of the male, give the breeders of this variety an opportunity of testing their skill in mating that is equaled in but few varieties of Standard fowls. The Columbians with their white breasts, backs and wing bows sharply contrasting with the black markings of necks and tails, present also an attractive color scheme.

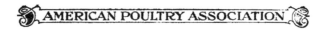

PLYMOUTH ROCKS.

Disqualifications.

Positive enamel white in ear-lobes. (See general disqualifications.)

STANDARD WEIGHTS.

Cock ....9½ lbs.   Hen ....7½ lbs.
Cockerel ....8 lbs.   Pullet ....6 lbs.

SHAPE OF MALE.

HEAD: Moderately large.

BEAK: Stout, comparatively short, regularly curved.

EYES: Full, prominent.

COMB: Single, rather small in proportion to size of specimen; set firmly on head; straight, upright; evenly serrated, having five well defined points, those in front and at rear a trifle smaller than the other three, giving the comb a semi-oval appearance when viewed from the side; fine in texture; blade not conforming too closely to head.

WATTLES AND EAR-LOBES: Wattles, moderately long, nicely rounded at the lower edges, equal in length, fine in texture, free from folds or wrinkles. Ear-lobes, oblong, smooth, hanging about one-third the length of wattles.

NECK: Rather long, slightly arched, having abundant hackle flowing well over shoulders.

WINGS: Of medium size, well folded; fronts, well covered by breast feathers and points well covered by saddle feathers.

BACK: Rather long, broad its entire length, flat at shoulders, nearly horizontal from neck to saddle, where there is a slight concave sweep to tail; saddle feathers, rather long, abundant, filling well in front of tail.

TAIL: Of medium length, moderately well spread, carried at an angle of forty-five degrees above the horizontal (see illustration, figures 39 and 40), forming no apparent angle with the back; sickles, well curved, covering tops of main tail feathers, conforming to the general shape of the tail; smaller sickles and tail-coverts, of medium length, nicely curved and sufficiently abundant to almost hide the stiff feathers of the tail when viewed from front or side.

BREAST: Broad, full, moderately deep, well rounded.

BODY AND FLUFF: Body, rather long, broad, deep, full, straight, extending well forward, connecting with breast so as to make no break in outline; fluff, moderately full.

LEGS AND TOES: Thighs, large, of medium length, well covered with soft feathers; shanks of medium length, smooth, straight, stout, set well apart; toes, straight, of medium length, well spread.

SHAPE OF FEMALE.

HEAD: Moderately large, broad, medium in length.

BEAK: Comparatively short, regularly curved.

EYES: Full, prominent.

COMB: Single, small, proportional to size of specimen; set firmly on the head; straight, upright; evenly serrated, having five well defined points, those in front and at rear being somewhat smaller and shorter than the other three.

WATTLES AND EAR-LOBES: Wattles, small, well rounded, equal in length, fine in texture. Ear-lobes, oblong in shape, smooth.

NECK: Medium in length, nicely curved and tapering to head, where it is comparatively small; neck feathers moderately full, flowing well over shoulders with no apparent break at juncture of neck and back.

WINGS: Of medium size, well folded; fronts, well covered by breast feathers.

BACK: Rather long, broad, its entire length, flat at shoulders, rising with a slightly concave incline to tail.

TAIL: Of medium length, fairly well spread, carried at an angle of thirty-five degrees above the horizontal (see illustration, fig. 39), forming no apparent angle with the back; tail-coverts, well developed.

BREAST: Broad, full, moderately deep, well rounded.

BODY AND FLUFF: Body, rather long, moderately deep, full, straight from front to rear and extending well forward, connected with the breast so as to make no break in outline; fluff, full, of medium length.

LEGS AND TOES: Thighs, of medium size and length, well covered with soft feathers; shanks, of medium length, set well apart, stout and smooth; toes, of medium size and length, straight, well spread.

BARRED PLYMOUTH ROCK MALE

BARRED PLYMOUTH ROCK FEMALE

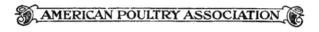

BARRED PLYMOUTH ROCKS.

Disqualifications.

Red in any part of plumage; two or more solid black primaries, secondaries or main tail feathers; shanks other than yellow, dark spots not to disqualify. (See general and Plymouth Rock disqualifications.)

COLOR OF MALE AND FEMALE.

BEAK: Yellow.

EYES: Reddish-bay.

COMB, FACE, WATTLES AND EAR-LOBES: Bright red.

SHANKS AND TOES: Yellow.

PLUMAGE: Grayish-white, each feather crossed by regular, narrow, parallel, sharply defined, dark bars that stop short of positive black; free from shafting, brownish tinge or metallic sheen; the light and dark bars to be of equal width, in number proportionate to length of feathers, and to extend throughout the length of feathers in all sections of the fowl; each feather ending with a narrow, dark tip; the combination of overlapping feathers giving the plumage a bluish appearance and of one even shade throughout.

WHITE PLYMOUTH ROCKS.

Disqualifications.

Red, buff or positive black in any part of plumage; shanks other than yellow. (See general and Plymouth Rock disqualifications.)

COLOR OF MALE AND FEMALE.

BEAK: Yellow.

EYES: Reddish-bay.

COMB, FACE, WATTLES AND EAR-LOBES: Bright red.

SHANKS AND TOES: Rich yellow.

PLUMAGE: Web, fluff and quills of feathers in all sections, pure white.

BUFF PLYMOUTH ROCKS.

Disqualifications.

Shanks other than yellow. (See general and Plymouth Rock disqualifications.)

COLOR OF MALE.

BEAK: Yellow.

EYES: Reddish-bay.

COMB, FACE, WATTLES AND EAR-LOBES: Bright red.

SHANKS AND TOES: Rich yellow.

PLUMAGE: Surface throughout an even shade of rich golden buff, free from shafting or mealy appearance, the head, neck, hackle, back, wing-bows and saddle richly glossed; under-color a lighter shade, free from foreign color. Different shades of buff in two or more sections is a serious defect. A harmonious blending of buff in all sections is most desirable.

COLOR OF FEMALE.

BEAK: Yellow.

EYES: Reddish-bay.

COMB, FACE, WATTLES AND EAR-LOBES: Bright red.

SHANKS AND TOES: Rich yellow.

PLUMAGE: Surface throughout an even shade of rich, golden buff, free from shafting or mealy appearance, the head and neck plumage showing a luster of the same shade as the rest of the plumage; under-color, a lighter shade, free from foreign color. Different shades of buff in two or more sections is a serious defect. A harmonious blending of buff in all sections is most desirable.

WHITE PLYMOUTH ROCK MALE

WHITE PLYMOUTH ROCK FEMALE

BUFF PLYMOUTH ROCK MALE

BUFF PLYMOUTH ROCK FEMALE

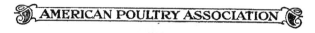

SILVER PENCILED PLYMOUTH ROCKS.

Disqualifications.

Shanks and toes other than yellow or dusky yellow. (See general and Plymouth Rock disqualifications.)

COLOR OF MALE.

HEAD: Plumage, silvery white.

BEAK: Yellow or dusky yellow.

EYES: Reddish-bay.

COMB, FACE, WATTLES AND EAR-LOBES: Bright red.

NECK: Hackle, web of feather, solid, lustrous greenish-black, with a narrow edging of silvery white, uniform in width, extending around point of feather; shafts, black; plumage in front of hackle, black.

WINGS: Bows, silvery white; coverts, lustrous greenish-black, forming a well defined bar of this color across wings when folded; primaries, black except a narrow edging of white on lower edge of lower webs; secondaries, black, except lower half of lower webs which should be white, except near end of feathers at which points the white terminates abruptly leaving end of feathers black.

BACK: Silvery white, free from brown; saddle, silvery white, with a black stripe in each feather, tapering to a point near its lower extremity.

TAIL: Black; sickles and coverts, lustrous greenish-black; smaller coverts, lustrous greenish-black edged with white.

BREAST: Black.

BODY AND FLUFF: Body, black; fluff, black slightly tinged with gray.

LEGS AND TOES: Thighs, black; shanks and toes, yellow or dusky yellow.

UNDER-COLOR OF ALL SECTIONS: Slate.

COLOR OF FEMALE.

HEAD: Plumage, silvery gray.

BEAK: Yellow or dusky yellow.

EYES: Reddish-bay.

COMB, FACE, WATTLES AND EAR-LOBES: Bright red.

NECK: Silvery white; center portion of feathers, black slightly penciled with gray; feathers in front of neck, same as breast.

WINGS: Shoulders, bows and coverts; gray with distinct dark pencilings, outlines of which conform to shape of feathers; primaries, black with narrow edge of gray penciling on lower webs; secondaries, upper webs, black, lower webs, gray with distinct dark pencilings extending around outer edge of feathers.

BACK: Gray, with distinct dark pencilings, outlines of which conform to shape of feather; feathers, free from white shafting.

TAIL: Black, except the two top feathers, which are penciled on upper edge; coverts, gray, with distinct dark pencilings, outlines of which conform to shape of feather.

BREAST: Gray, with distinct dark pencilings, outlines of which conform to shape of feather.

BODY AND FLUFF: Body, gray, with distinct dark pencilings, reaching well down on thighs; fluff, gray, penciled with a darker shade.

LEGS AND TOES: Thighs, gray, with distinct pencilings; shanks and toes, yellow or dusky yellow.

UNDER-COLOR OF ALL SECTIONS: Slate.

Note: Each feather in back, breast, body, wing-bows, and thighs to have three or more distinct pencilings.

57

SILVER PENCILED PLYMOUTH ROCK MALE

SILVER PENCILED PLYMOUTH ROCK FEMALE

PARTRIDGE PLYMOUTH ROCKS.

Disqualifications.

Positive white in main tail feathers, sickles or secondaries; shanks other than yellow or dusky yellow. (See general and Plymouth Rock disqualifications.)

COLOR OF MALE.

HEAD: Plumage, bright red.

BEAK: Dark horn, shading to yellow at point.

EYES: Reddish-bay.

COMB, FACE, WATTLES AND EAR-LOBES: Bright red.

NECK: Hackle, web of feather solid, lustrous greenish-black, with a narrow edging of rich, brilliant red, uniform in width, extending around point of feather; shaft, black; plumage in front of hackle, black.

WINGS: Fronts, black; bow, rich, brilliant red; coverts, lustrous greenish-black, forming a well defined bar of this color across wings when folded; primaries, black, lower edges, reddish bay; secondaries, black, outside webs, reddish bay, terminating with greenish-black at end of each feather.

BACK: Rich, brilliant red with lustrous greenish-black stripe down the middle of each feather, same as in hackle.

TAIL: Black; sickles and smaller sickles, lustrous greenish-black; coverts. lustrous greenish-black, edged with rich, brilliant red.

BREAST: Lustrous black.

BODY AND FLUFF: Body, black; fluff, black, slightly tinged with red.

LEGS AND TOES: Thighs, black; shanks and toes, yellow.

UNDER-COLOR OF ALL SECTIONS: Slate.

COLOR OF FEMALE.

HEAD: Plumage, mahogany-brown.

BEAK: Dark horn shading to yellow at point.

EYES: Reddish-bay.

COMB, FACE, WATTLES AND EAR-LOBES: Bright red.

NECK: Reddish-bay; center portion of feathers black, slightly penciled with mahogany-brown, feathers in front of neck, same as breast.

WINGS: Shoulders, bows and coverts, mahogany-brown, penciled with black, outlines of pencilings conforming to shape of feathers; primaries, black with edging of mahogany-brown on outer webs; secondaries, inner webs, black, outer webs mahogany-brown, penciled with black, outlines of pencilings conforming to shape of feathers.

BACK: Mahogany-brown, distinctly penciled with black, the outlines of pencilings conforming to shape of feathers.

TAIL: Black, the two top feathers penciled with mahogany-brown on upper edge; coverts, mahogany-brown penciled with black.

BREAST: Mahogany-brown. distinctly penciled with black, the outlines of pencilings conforming to shape of feathers.

BODY AND FLUFF: Body, mahogany-brown, penciled with black; fluff, mahogany-brown.

LEGS AND TOES: Thighs, mahogany-brown, penciled with black; shanks and toes, yellow or dusky yellow.

UNDER-COLOR OF ALL SECTIONS: Slate.

Note: Each feather in back, breast, body, wing-bows, and thighs to have three or more distinct pencilings.

PARTRIDGE PLYMOUTH ROCK MALE

PARTRIDGE PLYMOUTH ROCK FEMALE

COLUMBIAN PLYMOUTH ROCKS.

Disqualifications.

One or more solid black or brown feathers on surface of back of females; positive black spots prevalent in web of feathers of back except slight dark or black stripes in saddle near tail of males or in cape of either sex; red feathers in plumage; shanks other than yellow. (See general and Plymouth Rock disqualifications.)

COLOR OF MALE.

HEAD: Plumage, white.

BEAK: Yellow, with dark stripe down upper mandible.

EYES: Reddish-bay.

COMB, FACE, WATTLES AND EAR-LOBES: Bright red.

NECK: Hackle, web of feather solid, lustrous greenish-black with a narrow edging of white, uniform in width, extending around point of feather; greater portion of shaft, black; plumage in front of hackle, white.

WINGS: Bows, white except fronts, which may be partly black; coverts, white; primaries, black, with white edging on lower edge of lower webs; secondaries, lower portion of lower webs, white, sufficient to secure a white wing-bay, the white extending around ends of feathers and lacing upper portion of upper webs, this color growing wider in the shorter secondaries, sufficient to show white on surface when wing is folded; remainder of each secondary, black.

BACK: Surface color, white; cape, black and white; saddle, white, except feathers covering root and sides of tail which should be white with a narrow V-shaped black stripe at end of each feather tapering to a point near its lower extremity.

TAIL: Black; the curling feathers underneath, black laced with white; sickles and coverts, lustrous greenish-black; smaller coverts, lustrous greenish-black edged with white.

BREAST: Surface, white; under-color bluish-white, at juncture with body, bluish-slate.

BODY AND FLUFF: Body, white, except under wings, where it may be bluish-white; fluff, white.

LEGS AND TOES: Thighs, white; shanks and toes, yellow.

UNDER-COLOR OF ALL SECTIONS EXCEPT BREAST: Bluish-slate.

COLOR OF FEMALE.

HEAD: Plumage, white.

BEAK: Yellow, with dark stripe down upper mandible.

EYES: Reddish-bay.

COMB, FACE, WATTLES AND EAR-LOBES: Bright red.

NECK: Feathers beginning at juncture of head, web, a broad, solid lustrous greenish-black, with a narrow lacing of white extending around the outer edge of each feather; greater portion of shaft, black; feathers in front of neck, white.

WINGS: Bows, white; coverts, white; primaries, black, with white edging on lower edge of lower webs; secondaries, lower portion of lower webs, white, sufficient to secure a white wing-bay, the white extending around the ends and lacing upper portion of upper webs, this color growing wider in the shorter secondaries, sufficient to show white on surface when wing is folded; remainder of each secondary, black.

BACK: White; cape, black and white.

TAIL: Black, except the two top feathers which are laced with white; coverts, black with a narrow lacing of white.

BREAST: Surface, white; under-color, bluish-white, at juncture of body, bluish-slate.

BODY AND FLUFF: Body, white, except under wings where it may be bluish-white; fluff, white.

LEGS AND TOES: Thighs, white; shanks and toes, yellow.

UNDER-COLOR OF ALL SECTIONS EXCEPT BREAST: Bluish-slate.

COLUMBIAN PLYMOUTH ROCK MALE

COLUMBIAN PLYMOUTH ROCK FEMALE

WYANDOTTES.

The Wyandottes are of American origin and were known in their early history by several names. Each section of the country where they were found seems to have had a name that was given by the breeder who first introduced them. They were known as Sebrights, Mooneys, American Sebrights and by a number of names which their peculiar markings indicated. The name "Wyandotte" was not applied until they were admitted to the Standard in 1883. Just what breeds entered into the first Silver Wyandottes it is impossible to say. That Dark Brahmas and Silver Spangled Hamburgs were two of them has been proven, as a cross of these two breeds produces fowls that resemble them, but fail in shape and partly in color, showing that some other—[unknown] cross was added. They have, since their admission to the Standard, been one of the popular middle-weight breeds.

In shape the Wyandotte has a type peculiarly its own. It is emphatically a bird of curves. Breeders should strive to maintain the short, broad back and deep, round body; also, the curved, close-fitting comb which adds to the beauty of the specimen.

The wide range of color found in the eight varieties allows every admirer to indulge his fancy. Each variety has points of color difficult to obtain, but, when obtained, places a high valuation on the specimen. Whichever variety one may choose, he will find interesting color problems to solve. In the Whites, it will be how to secure pure white plumage and escape creaminess and brassiness: in the Blacks, how to obtain glossy greenish black, without the purple barring: in the Silvers, how to obtain silvery hackles and saddles free from brassiness; large, oval, white centers free from mossiness; and breast lacings free from white edgings: in the Goldens, how to get the correct shade of golden-bay, which in this variety supplants the white of the Silvers: in the Buffs, how to secure an even shade of rich, golden buff, and to avoid the out-cropping of black and white: in the Partridge and Silver Penciled varieties, how to obtain the rich foundation color with distinct, clean-cut lacings in the necks and backs of males, with the fine triple penciling in females: in the Columbians, how to keep the surface of necks, backs and wing-bows of males free from brassiness and secure distinct lacings in necks, with black tails, laced coverts and black and white wings in males and females.

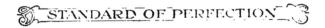

Wyandotte Disqualifications.

Ear-lobes more than one-quarter positive enamel white.

STANDARD WEIGHTS.

Cock ....8½ lbs. Hen ....6½ lbs.
Cockerel ....7½ lbs. Pullet ....5½ lbs.

SHAPE OF MALE.

HEAD: Short, round, broad.

BEAK: Short, well curved.

EYES: Full, oval.

COMB: Rose, low, firm on head; top, free from hollow center, oval, and surface covered with small, rounded points, tapering to a well defined point at rear; the entire comb curving to conform to the shape of skull.

WATTLES AND EAR LOBES: Wattles, moderately long, nicely rounded at lower edges, equal in length, fine in texture, free from folds or wrinkles. Ear-lobes, oblong, well defined, hanging about one-third the length of wattles; smooth.

NECK: Short, well arched; hackle, abundant, flowing well over shoulders.

WINGS: Medium in size, not carried too closely to body; sides, well rounded.

BACK: Short, broad, flat at shoulders; saddle, broad, full, rising with concave sweep to tail; saddle feathers, abundant.

TAIL: Short, well spread at base, carried at an angle of fifty degrees above the horizontal (see illustration, fig. 41); sickles, moderately long, curving gracefully and closely over tail; coverts, abundant, filling out well in front, hiding the stiff feathers.

BREAST: Broad, deep, round.

BODY AND FLUFF: Body, moderately short, deep round; fluff, full-feathered, well rounded.

LEGS AND TOES: Thighs, short, stout, showing outlines when viewed sideways, well covered with short feathers; shanks, short, stout, set well apart, well rounded; toes, straight.

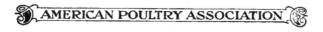

SHAPE OF FEMALE.

HEAD: Short, round; crown, broad.

BEAK: Short, well curved.

EYES: Full, oval.

COMB: Rose, similar to that of male, but much smaller.

WATTLES AND EAR-LOBES: Wattles, fine in texture, well rounded. Ear-lobes, oblong in shape, well defined.

NECK: Short, well arched; neck feathers, abundant.

WINGS: Medium in size, well rounded and well folded; fronts, well covered by breast feathers.

BACK: Short, broad, flat at shoulders; rising in a concave sweep to a broad, slightly rounded cushion, which extends well on to main tail; plumage, abundant.

TAIL: Short, well spread at base, carried at an angle of forty degrees above the horizontal (see illustration, fig. 39); coverts, abundant.

BREAST: Broad, deep, round.

BODY AND FLUFF: Body moderately short, deep, round; fluff, full-feathered, well rounded.

LEGS AND TOES: Thighs, short, stout, well spread, showing outlines when viewed sideways, well covered with soft feathers; shanks, short, stout, set well apart, well rounded; toes, straight.

SILVER WYANDOTTES.

Disqualifications.

Shanks other than yellow. (See general and Wyandotte disqualifications.)

COLOR OF MALE.

HEAD: Plumage, silvery white, each feather having a black stripe tapering to a fine point near its extremity.

BEAK: Dark horn, shading to yellow at point.

EYES: Reddish-bay.

COMB, FACE, WATTLES AND EAR-LOBES: Bright red.

NECK: Hackle, web of feather lustrous greenish-black with a narrow edging of silvery white, uniform in width, extending around point of feather; shaft of feather, white; plumage in front of hackle same as breast.

WINGS: Bows, silvery white; coverts, white with narrow lustrous greenish-black lacings, conforming to the shape of feathers, forming a double bar of laced feathers across wings; primaries, black, lower edges, white; secondaries, black, lower half of outer webs, white with narrow black edgings wider at the tips, upper webs, edged with white.

BACK: Silvery white; saddle, silvery white in appearance, a black stripe through each feather, laced with white, conforming to shape of center; the black having a long diamond-shaped, center of white.

TAIL: Black; sickles and coverts, lustrous greenish-black; smaller coverts, black, with diamond-shaped white centers, feathers laced with white.

BREAST: Web of each feather, white, laced with a narrow, lustrous greenish-black, sharply defined lacing, conforming to edge of feather.

BODY AND FLUFF: Body, web of each feather, white, laced with a narrow, lustrous greenish-black, sharply defined lacing, conforming to edge of feather; fluff, slate, powdered with gray.

LEGS AND TOES: Thighs, web of each feather, white, laced with a narrow, lustrous greenish-black, sharply defined lacing, conforming to edge of feather; shanks and toes, yellow.

UNDER-COLOR OF ALL SECTIONS: Slate.

COLOR OF FEMALE.

HEAD: Plumage, silvery gray.

BEAK: Dark horn, shading to yellow at point.

EYES: Reddish-bay.

COMB, FACE, WATTLES AND EAR-LOBES: Bright red.

NECK: Silvery white in appearance, with a black center through each feather, laced with white; shafts of feathers, white; feathers in front of neck same as breast.

WINGS: Shoulders, bows and coverts, each feather white, laced with a narrow, lustrous greenish-black, sharply defined lacing conforming to edge of feather; primaries, black, lower edges white; secondaries, black, lower half of outer webs, white with narrow black edging wider at tips.

SILVER WYANDOTTE MALE

SILVER WYANDOTTE FEMALE

BACK: Each feather white, laced with a narrow, lustrous greenish-black, sharply defined lacing, to conform to edge of feather.

TAIL: Black, the upper sides of the two top feathers edged with white; coverts, and smaller coverts, black with white centers.

BREAST: Each feather white, laced with a narrow, lustrous greenish-black, sharply defined lacing to conform to edge of feather.

BODY AND FLUFF: Body, each feather white, laced with a narrow, lustrous, greenish-black, sharply defined lacing to conform to edge of feather; fluff, slate powdered with gray.

LEGS AND TOES: Thighs, each feather white, laced with a narrow, lustrous, greenish-black, sharply defined lacing, to conform to edge of feather; shanks and toes yellow.

UNDER-COLOR OF ALL SECTIONS: Slate.

GOLDEN WYANDOTTES.

Disqualifications.

Shanks other than yellow or dusky yellow. (See general and Wyandotte disqualifications.)

COLOR OF MALE.

HEAD: Plumage, golden-bay, each feather having a black stripe, tapering to a fine point near its extremity.

BEAK: Dark horn, shading to yellow at point.

EYES: Reddish-bay.

COMB, FACE, WATTLES AND EAR-LOBES: Bright red.

NECK: Hackle, web of feather, lustrous greenish-black, with a narrow edging of golden-bay, uniform in width, extending around point of feather; shaft of feather, golden-bay; plumage in front of hackle, same as breast.

WINGS: Bows, golden bay; coverts, golden bay with narrow, lustrous greenish-black lacings, conforming to shape of feathers forming a double bar of laced feathers across the wings; primaries, black, lower edge, golden-bay; secondaries, black, lower half of outer webs, golden-bay, with a narrow black edging wider at the tip, upper webs, edged with golden-bay.

BACK: Golden-bay; saddle, golden-bay in appearance, a black stripe through each feather, laced with golden-bay, conforming to shape of center, the black having a long, diamond-shaped center of golden-bay.

TAIL: Black; sickles and coverts, lustrous greenish-black; smaller coverts, black, with diamond-shaped, golden-bay centers, feathers laced with golden-bay.

BREAST: Web of each feather, golden-bay, laced with a narrow, lustrous, greenish-black, sharply defined lacing, conforming to edge of feather.

BODY AND FLUFF: Body, web of each feather, golden-bay, laced with a narrow, lustrous, greenish-black, sharply defined lacing, conforming to edge of feather; fluff, slate, powdered with golden-bay.

LEGS AND TOES: Thighs, web of each feather, golden-bay, laced with a narrow, lustrous, greenish-black, sharply defined lacing, to conform to edge of feather; shanks and toes, yellow.

UNDER-COLOR OF ALL SECTIONS: Slate.

COLOR OF FEMALE.

HEAD: Plumage, golden-bay.

BEAK: Dark horn, shading to yellow at point.

EYES: Reddish-bay.

COMB, FACE, WATTLES AND EAR-LOBES: Bright red.

NECK: Golden-bay in appearance, with a black center through each feather, laced with golden-bay; shafts of feathers, golden-bay; feathers in front of neck same as breast.

WINGS: Shoulders, bows and coverts, each feather golden-bay laced with a narrow, lustrous greenish-black, sharply defined lacing, to conform to edge of feather; primaries, black, lower edges, golden-bay; secondaries, black, lower half of outer webs, golden-bay with narrow, black edgings wider at tips.

BACK: Each feather golden-bay, laced with a narrow lustrous greenish-black, sharply defined lacing, to conform to edge of feather.

TAIL: Black, the upper sides of the two top feathers edged with golden-bay; coverts and smaller coverts black with golden-bay center.

BREAST: Each feather, golden-bay, laced with a narrow, lustrous, greenish-black, sharply defined lacing, to conform to edge of feather.

BODY AND FLUFF: Body, each feather golden-bay, laced with a narrow, lustrous, greenish-black, sharply defined lacing, to conform to edge of feather; fluff, slate, powdered with golden-bay.

LEGS AND TOES: Thighs, each feather golden-bay, laced with a narrow, lustrous greenish-black, sharply defined lacing, to conform to edge of feather; shanks and toes, yellow.

UNDER-COLOR OF ALL SECTIONS: Slate.

WHITE WYANDOTTE MALE

76

WHITE WYANDOTTE FEMALE

WHITE WYANDOTTES.

Disqualifications.

Red, buff or positive black in any part of plumage; shanks other than yellow. (See general and Wyandotte disqualifications.)

COLOR OF MALE AND FEMALE.

BEAK: Yellow.
EYES: Reddish-bay.
COMB, FACE, WATTLES AND EAR-LOBES: Bright red.
SHANKS AND TOES: Rich yellow.
PLUMAGE: Web, fluff and quills of feathers in all sections, pure white.

BLACK WYANDOTTES.

Disqualifications.

Red in any part of plumage or white in any feather extending more than one-half inch; shanks other than black shading into yellow or dusky yellow; bottoms of feet other than yellow. (See general and Wyandotte disqualifications.)

COLOR OF MALE AND FEMALE.

BEAK: Black, shaded with yellow.
EYES: Reddish-bay.
COMB, FACE, WATTLES AND EAR-LOBES: Bright red.
SHANKS AND TOES: Yellow or dusky yellow.
BOTTOMS OF FEET: Yellow.
PLUMAGE: Lustrous greenish-black throughout.
UNDER-COLOR OF ALL SECTIONS: Slate.

BUFF WYANDOTTES.

Disqualifications.

Shanks other than yellow. (See general and Wyandotte disqualifications.)

COLOR OF MALE.

BEAK: Yellow.

EYES: Reddish-bay.

COMB, FACE, WATTLES AND EAR-LOBES: Bright red.

SHANKS AND TOES: Rich yellow.

PLUMAGE: Surface throughout, an even shade of rich, golden buff, free from shafting or mealy appearance; the head, neck, hackle, back, wing-bows and saddle, richly glossed; under-color, a lighter shade, free from foreign color. Different shades of buff in two or more sections is a serious defect. A harmonious blending of buff in all sections is most desirable.

COLOR OF FEMALE.

BEAK: Yellow.

EYES: Reddish-bay.

COMB, FACE, WATTLES AND EAR-LOBES: Bright red.

SHANKS AND TOES: Rich yellow.

PLUMAGE: Surface throughout, an even shade of rich, golden buff, free from shafting or mealy appearance; the head and neck plumage showing a luster of the same shade as the rest of the plumage; under-color, a lighter shade, free from foreign color. Different shades of buff in two or more sections is a serious defect. A harmonious blending of buff in all sections is most desirable.

BUFF WYANDOTTE MALE

BUFF WYANDOTTE FEMALE

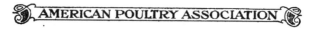

PARTRIDGE WYANDOTTES.

Disqualifications.

Positive white in main tail feathers, sickles or secondaries; shanks other than yellow or dusky yellow. (See general and Wyandotte disqualifications.)

COLOR OF MALE.

HEAD: Plumage, Bright-red.

BEAK: Dark horn, shading to yellow at point.

EYES: Reddish-bay.

COMB, FACE, WATTLES AND EAR-LOBES: Bright red.

NECK: Hackle, web of feather, solid, lustrous greenish-black with a narrow edging of rich, brilliant red, uniform in width, extending around point of feather; shaft, black; plumage in front of hackle, black.

WINGS: Fronts, black; bows, rich, brilliant red; coverts, lustrous greenish-black, forming a well defined bar of this color across wings when folded; primaries, black, lower edges, reddish-bay; secondaries, black, outside webs, reddish-bay, terminating with greenish-black at end of each feather.

BACK AND SADDLE: Rich, brilliant red, with lustrous greenish-black stripe down middle of each feather, same as in hackle.

TAIL: Black, sickles and smaller sickles, lustrous greenish-black; coverts, lustrous greenish-black edged with rich, brilliant red.

BREAST: Lustrous black.

BODY AND FLUFF: Body, black; fluff, black slightly tinged with red.

LEGS AND TOES: Thighs, black; shanks and toes, yellow.

UNDER-COLOR OF ALL SECTIONS: Slate.

COLOR OF FEMALE.

BEAK: Dark horn, shading to yellow at point.

EYES: Reddish-bay.

COMB, FACE, WATTLES AND EAR-LOBES: Bright red.

HEAD: Plumage, mahogany brown.

NECK: Reddish-bay, center portion of feathers, black slightly penciled with mahogany brown; feathers in front of neck same as breast.

WINGS: Shoulders, bows and coverts, mahogany brown, penciled with black, outlines of pencilings conforming to shape of feathers; primaries, black, with edging of mahogany brown on outer webs; secondaries, inner webs black, outer webs mahogany brown penciled with black, pencilings conforming to shape of feathers.

BACK: Mahogany brown, distinctly penciled with black, the outlines of pencilings conforming to shape of feathers.

TAIL: Black, the two top feathers black penciled with mahogany brown on upper edge; coverts, mahogany brown, penciled with black.

BREAST: Mahogany brown, distinctly penciled with black, the outlines of pencilings conforming to shape of feathers.

BODY AND FLUFF: Body, mahogany brown, penciled with black; fluff, mahogany brown.

SHANKS AND TOES: Yellow or dusky yellow.

UNDER-COLOR OF ALL SECTIONS: Slate.

Note: Each feather in back, breast, body, wing-bows and thighs to have three or more distinct pencilings.

SILVER PENCILED WYANDOTTES.

Disqualifications.

Shanks and toes, other than yellow or dusky yellow. (See general and Wyandotte disqualifications.)

COLOR OF MALE.

HEAD: Plumage, silvery white.

BEAK: Yellow or dusky yellow.

EYES: Reddish-bay.

COMB, FACE, WATTLES AND EAR-LOBES: Bright red.

NECK: Hackle, web of feather, solid, lustrous greenish-black, with a narrow edging of silvery white, uniform in width, extending around point of feather; shafts, black; plumage in front of hackle, black.

WINGS: Bows, silvery white; coverts, lustrous greenish-black; primaries, black, except a narrow edging of white on lower edge of lower webs; secondaries, black, except lower half of lower webs which should be white, except near end of feathers at which points the white terminates abruptly, leaving end of feathers black.

PARTRIDGE WYANDOTTE MALE

PARTRIDGE WYANDOTTE FEMALE

BACK: Silvery white, free from brown; saddle, silvery white with a black stripe in each feather, tapering to a point near its lower extremity.

TAIL: Black; sickles and coverts, lustrous greenish-black; smaller coverts, lustrous greenish-black, edged with white.

BREAST: Black.

BODY AND FLUFF: Body, black; fluff, black slightly tinged with gray.

LEGS AND TOES: Thighs, black; shanks and toes, yellow or dusky yellow.

UNDER-COLOR OF ALL SECTIONS: Slate.

COLOR OF FEMALE.

HEAD: Plumage, silvery gray.

BEAK: Yellow or dusky yellow.

EYES: Reddish-bay.

COMB, FACE, WATTLES AND EAR-LOBES: Bright red.

NECK: Silvery white, center portion of feathers, black slightly penciled with gray; feathers in front of neck same as breast.

WINGS: Shoulders, bows and coverts, gray with distinct dark pencilings, outlines of which conform to shape of feathers; primaries, black with a narrow edge of gray penciling on lower webs; secondaries, upper webs, black, lower webs, gray, with distinct dark pencilings extending around outer end of feathers.

BACK: Gray with distinct dark pencilings, outlines of which conform to shape of feathers; feathers free from white shafting.

TAIL: Black, except the two top feathers, which are penciled on upper edge; coverts, gray with distinct dark pencilings, outlines of which conform to shape of feathers.

BREAST: Gray with distinct dark pencilings, outlines of which conform to shape of feathers.

BODY AND FLUFF: Gray with distinct dark pencilings, reaching well down on thighs; fluff, gray, penciled with a darker shade.

LEGS AND TOES: Thighs, gray with distinct pencilings, shanks and toes, yellow, or dusky yellow.

UNDER-COLOR OF ALL SECTIONS: Slate.

Note: Each feather in back, breast, body, wing-bows and thighs to have three or more distinct pencilings.

COLUMBIAN WYANDOTTES.

Disqualifications.

One or more solid black or brown feathers on surface of back of female; positive black spots prevalent in web of back, except slight dark or black stripes in saddle near tail of male, or in cape of either sex; shanks other than yellow. (See general and Wyandotte disqualifications.)

COLOR OF MALE.

HEAD: Plumage, white.

BEAK: Yellow, with dark stripe down upper mandible.

EYES: Reddish-bay.

COMB, FACE, WATTLES AND EAR-LOBES: Bright red.

NECK: Hackle, web of feather, solid, lustrous greenish-black, with a narrow edging of white, uniform in width, extending around point of feather; greater portion of shaft, black; plumage in front of hackle, white.

WINGS: Bows and coverts, white, except fronts, which may be partly black; primaries, black, with white edging on lower edge of lower webs; secondaries, lower portion of lower webs, white, sufficient to secure a white wing-bay, the white extending around end of feathers, and lacing upper portion of upper webs, this color growing wider in the shorter secondaries, sufficient to show white on surface when wing is folded; remainder of each secondary, black.

BACK: Surface color, white; cape, black and white; saddle, white, except feathers covering root and sides of tail which should be white with a narrow V-shaped, black stripe at end of each feather, tapering to a point near its lower extremity.

TAIL: Black; the curling feathers underneath, black laced with white; sickles and coverts, lustrous greenish-black; smaller coverts, lustrous greenish-black edged with white.

BREAST: Surface, white; undercolor, bluish-white, at juncture with body, bluish-slate.

BODY AND FLUFF: Body, white, except under wings, where it may be bluish-white; fluff, white.

LEGS AND TOES: Thighs, white; shanks and toes, yellow.

UNDER-COLOR OF ALL SECTIONS EXCEPT BREAST: Bluish-slate.

COLUMBIAN WYANDOTTE MALE

COLUMBIAN WYANDOTTE FEMALE

COLOR OF FEMALE.

HEAD: Plumage, white.

BEAK: Yellow, with dark stripe down upper mandible.

EYES: Reddish-bay.

COMB, FACE, WATTLES AND EAR-LOBES: Bright red.

NECK: Feathers beginning at juncture of head, web, a broad, solid, lustrous greenish-black, with a narrow lacing of white extending around the outer edge of each feather; shaft, black; feathers in front of neck, white.

WINGS: Bows and coverts, white; primaries, black with white edging on lower edge of lower webs; secondaries, lower portion of lower webs, white, sufficient to secure a white wing-bay, the white extending around the end and lacing upper portion of upper webs, this color growing wider in the shorter secondaries, sufficient to show white on surface when wing is folded; remainder of each secondary, black.

BACK: White; cape, black and white.

TAIL: Black, except the two top feathers, which are laced with white; coverts, black, with a narrow lacing of white.

BREAST: Surface, white; under-color bluish-white, at juncture with body, bluish-slate.

BODY AND FLUFF: Body, white, except under wings, where it may be bluish-white; fluff, white.

LEGS AND TOES: Thighs, white; shanks and toes, yellow.

UNDER-COLOR OF ALL SECTIONS EXCEPT BREAST: Bluish-slate.

JAVAS.

This breed presents the extreme length of body found in breeds of the American class. The length of both back and body, together with the breadth of back, the depth of body, the full, well rounded breast and smooth posterior, give the breed a type peculiarly its own. The color of plumage of the black variety is a rich, lustrous black with greenish sheen. Purple barring is a serious defect. The color of plumage of the mottled variety is black and white throughout. These colors should be sharply divided, each distinct in itself, the black predominating.

STANDARD WEIGHTS.

Cock ....9½ lbs.   Hen ....7½ lbs.
Cockerel 8   lbs.   Pullet ....6½ lbs.

SHAPE OF MALE.

HEAD: Of medium length, and breadth.

BEAK: Stout, well curved.

EYES: Large, full.

COMB: Single, rather small, straight and upright; firm on head; lower in front; evenly serrated having five well defined points; fine in texture.

WATTLES AND EAR-LOBES: Wattles, of medium length, well rounded at ends, smooth, fine in texture. Ear-lobes, small, oblong.

NECK: Of medium length, arched; hackle, abundant.

WINGS: Rather large, well folded.

BACK: Broad, long, with a slight decline to a concave sweep near tail; saddle feathers, abundant.

TAIL: Rather long, moderately full and expanded, carried at an angle of forty-five degrees above the horizontal (see illustration, fig. 40), sickles, long and gracefully curved; main tail feathers, long.

BREAST: Broad, full, deep.

BODY AND FLUFF: Body, long, broad, deep; fluff, moderately full.

LEGS AND TOES: Thighs, of medium length, large, strong, well covered with close-fitting feathers; shanks, of medium length; stout in bone; toes, of medium length, straight, strong, well spread.

BLACK JAVA MALE

BLACK JAVA FEMALE

SHAPE OF FEMALE.

HEAD: Of medium size.

BEAK: Strong, well curved.

EYES: Of medium size, oval, full.

COMB: Single, small, straight and upright, lower in front; evenly serrated, having five well-defined points; fine in texture.

WATTLES AND EAR-LOBES: Wattles, of medium size, well rounded, smooth, fine in texture. Ear-lobes, small.

NECK: Of medium length, slightly arched.

WINGS: Rather large, well folded.

BACK: Long, full near tail coverts.

TAIL: Rather long, full, slightly expanded, carried at an angle of forty-five degrees above the horizontal. (See illustration, fig. 39.)

BREAST: Broad, full, deep.

BODY AND FLUFF: Body, long, broad, deep; fluff, moderately full, even on surface.

LEGS AND TOES: Thighs, of medium length, large, strong, well covered with close-fitting feathers; shanks of medium length, stout in bone; toes, of medium length, straight, strong, well spread.

BLACK JAVAS.

Disqualifications.

Positive enamel white in ear-lobes; foreign color in any part of plumage; skin, or bottoms of feet, other than yellow. (See general disqualifications.)

COLOR OF MALE AND FEMALE.

BEAK: Black.

EYES: Black or dark brown.

COMB, FACE, WATTLES AND EAR-LOBES: Red or gipsy color.

SHANKS AND TOES: Shanks, black or nearly black, with a tendency toward willow, black preferred; toes, same as shanks, except under part which must be yellow, bottom of feet, yellow.

PLUMAGE: Rich, lustrous black with greenish sheen, free from purple barring.

UNDER-COLOR OF ALL SECTIONS: Dull black.

MOTTLED JAVAS.

Disqualifications.

Positive enamel white in ear-lobes; red or brassy color in any part of plumage; skin or bottoms of feet other than yellow. (See general disqualifications.)

COLOR OF MALE AND FEMALE.

BEAK: Horn, or horn and yellow.

EYES: Reddish-bay.

COMB, FACE, WATTLES AND EAR-LOBES: Red.

SHANKS AND TOES: Broken leaden-blue and yellow.

PLUMAGE: Mottled black and white throughout, black predominating.

UNDER-COLOR OF ALL SECTIONS: Slate.

DOMINIQUES.

Disqualifications.

Positive enamel white in ear-lobes; any feather or feathers, or portion of a feather, of any color foreign to the breed, excepting solid black or white feathers. (See general disqualifications.)

STANDARD WEIGHTS.

Cock ....7 lbs.   Hen ....5 lbs.
Cockerel ....6 lbs.   Pullet ....4 lbs.

SHAPE OF MALE.

HEAD: Of medium size, carried well up.

BEAK: Short, stout, well curved.

EYES: Large, oval.

COMB: Rose, not so large as to overhang the eyes or beak; firm and straight on head; square in front; uniform on sides; free from hollow center; terminating in a spike at rear, the point of which turns slightly upward; top covered with small points.

WATTLES AND EAR-LOBES: Wattles, broad, full, pendant. Ear-lobes, oblong, of medium size.

NECK: Of medium length, well arched, tapering; hackle, abundant.

WINGS: Rather large, well folded; wing-bows and points, well covered by breast and saddle feathers.

BACK: Of medium length, broad, rising with concave sweep to tail.

DOMINIQUE MALE

DOMINIQUE FEMALE

TAIL: Long, full, slightly expanded; carried at an angle of forty-five degrees above the horizontal, (see illustrations, figures 39 and 40); sickles, long, well curved.

BREAST: Broad, round and carried well up.

BODY AND FLUFF: Body, broad, full, compact; fluff, moderately full.

LEGS AND TOES: Thighs, of medium length, strong, well covered with soft feathers; shanks, fine in bone; toes, of medium length, straight, well spread.

SHAPE OF FEMALE.

HEAD: Small.

BEAK: Short, stout, regularly curved.

EYES: Large, oval.

COMB: Rose, similar to that of male, but much smaller.

WATTLES AND EAR-LOBES: Wattles, rather small and well rounded. Ear-lobes, of medium size, oblong.

NECK: Short, slightly arched, tapering.

WINGS: Rather large, well folded.

BACK: Of medium length, broad, slightly concave.

TAIL: Full, rather long, slightly expanded, carried at an angle of forty-five degrees above the horizontal. (See illustrations, figures 39 and 40.)

BREAST: Round, full.

BODY AND FLUFF: Body, broad, full, compact; fluff, moderately full.

LEGS AND TOES: Thighs, of medium length, strong, well covered with soft feathers; shanks, fine in bone; toes, of medium length, straight, well spread.

COLOR OF MALE AND FEMALE.

BEAK: Yellow.

EYES: Reddish-bay.

COMB, FACE, WATTLES AND EAR-LOBES: Bright red.

SHANKS AND TOES: Yellow.

PLUMAGE: Slate; feathers in all sections of fowl crossed throughout their entire length by irregular dark and light bars that stop short of positive black and white; tip of each feather, dark; free from shafting, brownish tinge or metallic sheen; excellence to be determined by distinct contrasts. The male may be one or two shades lighter than the female.

UNDER-COLOR OF ALL SECTIONS: Slate.

RHODE ISLAND REDS.

(Single and Rose Combs.)

Rhode Island Reds are extensively bred and recognized as one of the most beautiful and useful of standard-bred fowls.

They are an American production which have been bred in large numbers for practical purposes in Rhode Island, Massachusetts, and adjoining states for about half a century, taking the name of Rhode Island from that State when first admitted to the Standard. They are believed to have originated from crosses of the Asiatics, Mediterranean and Games. Their chief characteristics are oblong shape, compact form and red color.

Disqualifications.

Positive enamel white covering more than one-fourth of ear-lobes; one or more entirely white feathers showing in outer plumage; shanks and feet other than yellow or reddish-horn. (See general disqualifications.)

STANDARD WEIGHTS.

Cock ....8½ lbs.   Hens ....6½ lbs.
Cockerel ....7½ lbs.   Pullet 5   lbs.

SHAPE OF MALE.

HEAD: Medium size, carried horizontally and slightly forward.

BEAK: Medium length, slightly curved.

EYES: Large, oval, prominent.

COMB: Single; medium in size, set firmly on head, perfectly straight and upright, with five even and well defined points, those in front and rear smaller than those in center; of considerable breadth where it joins to head; blade, smooth, inclining slightly downward but not following too closely the shape of head.

COMB: Rose; low, firm on head; oval, free from hollow center, surface covered with small rounded points, terminating in a spike at the rear, the spike drooping slightly but not conforming too closely to the shape of head.

WATTLES AND EAR-LOBES: Wattles, of medium size, equal in length, free from folds and wrinkles. Ear-lobes, oblong, well defined, smooth, in size proportionate to other head adjuncts.

SINGLE COMB RHODE ISLAND RED MALE

SINGLE COMB RHODE ISLAND RED FEMALE

Neck: Of medium length; hackle, abundant, flowing over shoulders, not too closely feathered.

Wings: Of good size, well folded, carried horizontally.

Back: Broad, long, carried horizontally, with slight concave sweep to tail; saddle feathers, of medium length, abundant.

Tail: Of medium length, well spread, carried at an angle of forty degrees above the horizontal (see illustration, fig. 39), thus increasing the apparent length of the fowl; sickles, of medium length, extending slightly beyond main tail feathers; smaller sickles and tail-coverts, of medium length, well covered with soft feathers.

Breast: Deep, full, well rounded.

Body and Fluff: Body, broad, deep, long, straight, extending well forward giving body oblong appearance; feathers carried close to body; fluff, moderately full.

Shanks and Toes: Shanks, of medium length, well rounded, smooth, set well apart; toes, of medium length, straight well spread.

SHAPE OF FEMALE.

Head: Medium size, carried horizontally and slightly forward.

Beak: Medium length, slightly curved.

Eyes: Large, oval.

Comb: Single; medium in size, set firmly on head, perfectly straight and upright, with five even and well defined points, those in front and rear smaller than those in center.

Comb: Rose; low, free from hollow center, firm on head, much smaller than that of male and in proportion to its length, narrower; covered with small points and terminating in a small, short spike at the rear.

Wattles and Ear-Lobes: Wattles, of medium size, equal in length, regularly curved. Ear-lobes, oblong, well defined, smooth, proportionate in size to other head adjuncts.

Neck: Of medium length, moderately full.

Wings: Rather large, well folded; fronts, well covered by breast feathers; flights, carried nearly horizontally.

Back: Broad, long, carried horizontally.

Tail: Medium length, moderately spread, carried at an angle of thirty-five degrees above the horizontal. (See illustration, figure 39.)

Breast: Deep, full, well rounded.

Body and Fluff: Body, broad, deep, long, straight, extending well forward, giving body an oblong appearance; feathers, carried close to body; fluff, moderately full.

LEGS AND TOES: Thighs, of medium length, well covered with soft feathers; shanks, of medium length, well rounded, smooth; toes, of medium length, strong, straight, well spread.

COLOR OF MALE.

HEAD: Plumage, brilliant red.

BEAK: Reddish-horn.

EYES: Reddish-bay.

COMB, FACE, WATTLES AND EAR-LOBES: Bright red.

NECK: Rich, brilliant red; plumage in front of neck, rich red.

WINGS: Bows, rich, brilliant red; coverts, red; primaries, upper webs, red, lower webs, black with narrow edging of red, sufficient only to prevent the black from showing on surface when wings are folded in natural position; primary coverts, black edged with red; secondaries, lower webs, red, the red extending around end of feathers sufficient to secure a red wing-bay and lacing the upper portion of upper webs, this color growing wider in shorter secondaries; remainder of each secondary black, feathers next to body being red on surface so that wing when folded in natural position shall show one harmonious red color.

BACK: Rich, brilliant red.

TAIL: Main tail, black; sickle feathers, black or greenish-black; coverts, mainly black, red as they approach the saddle.

BREAST: Rich red.

BODY AND FLUFF: Rich red.

SHANKS AND TOES: Rich yellow, or reddish horn color. A line of red pigment down the sides of shanks, extending to the tip of toes is desirable.

PLUMAGE: General surface, rich brilliant red, except where black is specified; free from shafting or mealy appearance; the less contrast there is between wing-bows, back, hackle and breast, the better. A harmonious blending in all sections is desired. The specimen should be so brilliant in color as to have a glossed appearance.

UNDER-COLOR OF ALL SECTIONS: Red.

COLOR OF FEMALE.

HEAD: Plumage, brilliant red.

BEAK: Reddish-horn.

EYES: Reddish-bay.

COMB, FACE, WATTLES AND EAR-LOBES: Bright red.

103

ROSE COMB RHODE ISLAND RED MALE

ROSE COMB RHODE ISLAND RED FEMALE

NECK: Rich red, with slight ticking of black, confined to tips of lowest neck feathers; feathers in front of neck, rich red.

WINGS: Bows, rich red; coverts, red; primaries, upper webs red, lower webs, black with narrow edging of red, sufficient only to prevent the black from showing on surface when wings are folded in natural position; primary coverts, black edged with red; secondaries, lower webs red, the red extending around the end of the feathers sufficient to secure a red wing-bay and lacing the upper portion of upper webs, this color growing wider in the shorter secondaries; remainder of each secondary, black, feathers next to body being red on surface, so that wing when folded in natural position shall show one harmonious red color.

BACK: Rich red.

TAIL: Black; the two top feathers may be edged with red.

BREAST: Rich red.

BODY AND FLUFF: Red.

SHANKS AND TOES: Rich yellow, or reddish-horn.

PLUMAGE: General surface color, rich, even red, except where black is specified; free from shafting or mealy appearance.

UNDER-COLOR OF ALL SECTIONS: Red.

BUCKEYES.

These fowls are strictly an American production and originated in Ohio. Their blood lines come from the Cornish, Black Breasted Red Game, Buff Cochin and Barred Plymouth Rock. From the Cornish comes the pea-comb. In appearance they most resemble the Cornish, being strongly Oriental in type, the thighs stout and muscular, the breast broad and rounded at sides, showing lack of fullness.

Disqualifications.

Positive enamel white covering more than one-fourth of earlobes; entirely white feathers in plumage. (See general disqualifications.)

STANDARD WEIGHTS.

Cock ....9 lbs. Hen ....6½ lbs.
Cockerel ....8 lbs. Pullet ....5½ lbs.

SHAPE OF MALE.

HEAD: Of medium size, carried well up.

BEAK: Short, stout, regularly curved.

EYES: Of medium size, full, with bold expression.

COMB: Pea; of medium size; firm, set closely on head.

WATTLES AND EAR-LOBES: Wattles, short, of equal length, moderately rounded. Ear-lobes, of medium size.

NECK: Of medium length, well arched, tapering nicely; hackle, abundant, flowing well over shoulders.

WINGS: Of medium size, well folded; wing-fronts and wing-points, well covered by breast and saddle feathers.

BACK: Broad, long, sloping slightly downward to base of tail; saddle feathers, rather short.

TAIL: Of medium length and size, carried at an angle of forty-five degrees above the horizontal (see illustrations, figures 39 and 40); sickles and coverts, of medium length, nicely curved, sufficiently abundant to cover well the stiff feathers.

BREAST: Broad, deep, well rounded, carried somewhat elevated above the horizontal.

BODY AND FLUFF: Body, rather long, broad, deep, full, heavy for size of bird, extending well forward; fluff, moderately full.

LEGS AND TOES: Thighs, of medium length, large, well covered with soft feathers; shanks, of medium length, stout, smooth, set well apart; toes, of medium length, straight, strong, well spread.

SHAPE OF FEMALE.

HEAD: Of medium size, carried well up.

BEAK: Short, stout, regularly curved.

EYES: Of medium size, full.

COMB: Pea; small, set closely on head.

WATTLES AND EAR-LOBES: Wattles, short, of equal length, moderately rounded. Ear-lobes, of medium size.

NECK: Of medium length, well curved; neck feathers, moderately full.

WINGS: Of medium size, well folded.

BACK: Broad, long, sloping slightly downward to base of tail.

TAIL: Of medium length, fairly well spread, carried at an angle of forty degrees above the horizontal. (See illustration, fig. 39.)

BREAST: Broad, deep, well rounded, carried somewhat elevated above the horizontal.

BODY AND FLUFF: Body, long, broad, deep, full, heavy for size of bird, extending well forward; fluff, moderately full.

LEGS AND TOES: Thighs, of medium length and size, well covered with soft feathers; shanks, of medium length, stout, smooth, set well apart; toes, of medium length and size, straight, well spread.

BUCKEYE MALE

BUCKEYE FEMALE

COLOR OF MALE.

BEAK: Yellow, shaded with reddish-horn.

EYES: Reddish-bay.

COMB, FACE, WATTLES AND EAR-LOBES: Bright red.

SHANKS AND TOES: Yellow.

PLUMAGE: Surface, mahogany bay, slightly accentuated on wing-bows; the unexposed flight and main tail feathers may contain black; sickles and coverts, shaded bay and black, thus avoiding a sharp contrast between body and tail; shafts of feathers, bay entire length.

UNDER-COLOR OF ALL SECTIONS: Red, except back, where there should be a distinct bar of slate across the feathers below the surface.

COLOR OF FEMALE.

BEAK: Yellow.

EYES: Reddish-bay.

COMB, FACE, WATTLES AND EAR-LOBES: Bright red.

SHANKS AND TOES: Yellow.

PLUMAGE: Surface, mahogany bay; the unexposed flight and main tail feathers may contain black; shafts of feathers, bay entire length.

UNDER-COLOR OF ALL SECTIONS: Red, except back, where there should be a distinct bar of slate across the feathers below the surface.

ASIATIC.

Breeds	*Varieties*

BRAHMAS .... {Light / Dark}

COCHINS .... {Buff / Partridge / White / Black}

LANGSHANS .... {Black / White}

SCALE OF POINTS FOR ASIATIC CLASS.

Symmetry	4
Weight	4
Condition	4
Comb	8
Head—Shape 2, Color 2................................	4
Beak—Shape 2, Color 2................................	4
Eyes—Shape 2, Color 2................................	4
Wattles and Ear-Lobes—Shape 2, Color 2................	4
Neck—Shape 4, Color 6................................	10
Wings—Shape 4, Color 6...............................	10
Back—Shape 6, Color 4................................	10
Tail—Shape 5, Color 5................................	10
Breast—Shape 6, Color 4...............................	10
Body and Fluff—Shape 5, Color 3.......................	8
Legs and Toes—Shape 3, Color 3.......................	6
	100

BRAHMAS.

The Brahma male should have that strength and grace of carriage which naturally belongs to a well proportioned fowl of its size and finish. The head, when well furnished, adds style and character to the specimen's commanding appearance. The body should be large, well rounded and free from any tendency to excessive fluff. The Standard does not provide for apparent cushion in Brahmas, either on male or female. The Brahma male should be of a distinctive type, unlike the Cochin in form and feather, being more compactly and firmly put together. The solidity of form and compactness of plumage unite in the standard Brahma male, to produce a finely proportioned fowl of large size and active nature. Comb, color and markings should be well defined, embodying the true Brahma characteristics. The Brahma female has the fine, graceful lines that properly belong to her as the mate of the stately and powerful male of this breed. She lacks rotundity of form as compared with the full-feathered Cochin, her body being more compact and closely feathered. The proper sweep of back from saddle to tail is formed largely by the distinctively Brahma spread of tail, which continues and finishes the back line, and fills out the side lines to proper form, within the true contour of Brahma shape.

Brahma Disqualifications.

Vulture-like hocks; shanks other than yellow or reddish-yellow.

SHAPE OF MALE.

HEAD: Of medium length, broad; crown projecting well over eyes.

BEAK: Stout, well curved.

EYES: Large, deep set.

COMB: Pea; small, firm and even on head, lower and narrower in front and rear than at center; each row evenly serrated; points in front and rear smaller than those at center.

WATTLES AND EAR-LOBES: Wattles, of medium size, well rounded. Ear-lobes, large, the lower edges on a level with or slightly below edges of wattles.

NECK: Moderately long, well arched; hackle, abundant, flowing over shoulders and meeting under throat, dividing at lower ends of wattles and flowing full at sides.

WINGS: Small, carried rather high, with lower line nearly horizontal; sides, well rounded; primaries, closely folded under secondaries.

BACK: Broad, rather long, flat across shoulders, carrying its width well back to tail, rising with slightly concave incline from shoulders to middle of saddle, where it takes a more pronounced concave sweep well up on tail; saddle, abundant, flowing full over sides, filling well in front of tail and covering wing-points.

TAIL: Rather large, full, well spread, carried high enough to continue concave sweep of back, filled underneath with curling feathers; sickles, short, spreading laterally; coverts, plentiful, but not so long as to cover the entire length of lower main tail feathers.

BREAST: Broad, deep, well rounded.

BODY AND FLUFF: Body, rather long, deep well rounded at sides; fluff, abundant, smooth in surface, giving specimen a broad but compact appearance.

LEGS AND TOES: Legs, straight, set well apart; thighs, stout, well covered with soft feathers, nicely rounded, free from vulture-like feathering; shanks, large, stout in bone, of sufficient length to properly balance specimen, well covered on outer sides with feathers; toes, straight, stout; outer and middle toes, well feathered.

SHAPE OF FEMALE.

HEAD: Of medium length, broad; crown projecting well over eyes.

BEAK: Stout, well curved.

EYES: Large, deep set.

COMB: Pea; low, firm and even on head; well serrated, the middle row higher and more distinctly serrated than the other two.

WATTLES AND EAR-LOBES: Wattles, small. Ear-lobes, large.

NECK: Of medium length, slightly arched; rather full under throat, hens having dewlap between wattles.

WINGS: Small, carried rather high, with lower line nearly horizontal; sides, well rounded; fronts, covered by breast feathers; primaries, closely folded under secondaries.

BACK: Broad, rather long, flat across shoulders with moderate incline to tail, carrying the width well back on tail.

TAIL: Of medium length, well spread at base, resembling an inverted "V" with wide angle when viewed from rear; carried high enough to continue the sweep of the back; tail-coverts, two rows, covering a greater part of both sides of main tail.

BREAST: Deep, broad, well rounded.

BODY AND FLUFF: Body, long, deep, well rounded at sides; fluff, abundant, smooth in surface, giving specimen a broad but compact appearance.

LEGS AND TOES: Legs, straight, set well apart; thighs, stout, well covered with soft feathers; shanks, of medium length, stout in bone, well feathered on outer sides; toes, straight, stout; outer and middle toes, well feathered.

LIGHT BRAHMAS.

Disqualifications.

One or more solid black or brown feathers on surface of back of female; positive black spots prevalent in web of feathers of back, except slight dark or black stripes in saddle of male, near tail, or in cape of either sex. (See general and Brahma disqualifications.)

STANDARD WEIGHTS:

Cock ....12 lbs. Hen ....9½ lbs.
Cockerel ....10 lbs. Pullet 8   lbs.

COLOR OF MALE.

HEAD: Plumage, white.

BEAK: Yellow, with dark stripe down upper mandible.

EYES: Reddish-bay.

COMB, FACE, WATTLES AND EAR-LOBES: Bright red.

NECK: Hackle, web of feathers a solid lustrous greenish-black with a narrow edging of white, uniform in width, extending around point of feather; greater portion of shaft, black; plumage in front of hackle, white.

WINGS: Bows and coverts, white, except fronts which may be partly black; primaries, black with white edging on lower edge of lower webs; secondaries, lower portion of lower web, white, sufficient to secure a white wing-bay, the white extending around end of feathers and lacing upper portion of upper webs, this color growing wider in the shorter secondaries, sufficient to show white on surface when wing is folded; remainder of each secondary, black.

BACK: Surface color, white; cape, black and white; saddle,

white, except feathers covering root and sides of tail which should be white with a narrow "V" shaped black stripe at end of each feather tapering to a point near its lower extremity.

TAIL: Black; the curling feathers underneath, black laced with white; sickles and coverts, lustrous greenish-black; smaller coverts, lustrous greenish-black, edged with white.

BREAST: Surface, white; under-color, bluish-white, at juncture with body, bluish-slate.

BODY AND FLUFF: Body, white except under wings, where it may be bluish-white; fluff, white.

LEGS AND TOES: Thighs, white; shank feathers, white and black; outer toe feathering, white and black, where black, laced with white; shanks and toes, yellow.

UNDER-COLOR OF ALL SECTIONS EXCEPT BREAST: Bluish-slate.

COLOR OF FEMALE.

HEAD: Plumage, white.

BEAK: Yellow, with dark stripe down the upper mandible.

EYES: Reddish-bay.

COMB, FACE, WATTLES AND EAR-LOBES: Bright red.

NECK: Feathers beginning at juncture of head, web, a broad, solid, lustrous greenish-black with a narrow lacing of white extending around the outer edge of each feather; greater portion of shafts, black; feathers in front of neck, white.

WINGS: Bows and coverts, white; primaries, black with white edging on lower edge of lower webs; secondaries, lower portion of lower webs white, sufficient to secure a white wing-bay, the white extending around the end and lacing upper portion of upper webs, this color growing wider in the shorter secondaries, sufficient to show white on surface when wing is folded; remainder of each secondary, black.

BACK: White; cape, black and white.

TAIL: Black, except the two top feathers, which are laced with white; coverts, black, with a narrow lacing of white.

BREAST: Surface, white; under-color, bluish-white, at juncture of body, bluish-slate.

BODY AND FLUFF: Body, white, except under wings, where it may be bluish-white; fluff, white.

LEGS AND TOES: Thighs, white; shank feathers, white; outer toe feathers, white and black, where black, laced with white; shanks and toes, yellow.

UNDER-COLOR OF ALL SECTIONS EXCEPT BREAST: Bluish-slate.

LIGHT BRAHMA MALE

116

LIGHT BRAHMA FEMALE.

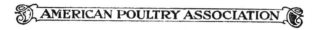

DARK BRAHMAS.

Disqualifications.

(See general and Brahma disqualifications.)

STANDARD WEIGHTS:

Cock ....11 lbs.   Hen ....8½ lbs.
Cockerels .... 9 lbs.   Pullet ....7 lbs.

COLOR OF MALE.

HEAD: Plumage silvery white.

BEAK: Dark horn shading to yellow at point.

EYES: Reddish-bay.

COMB, FACE, WATTLES AND EAR-LOBES: Bright red.

NECK: Hackle, web of feathers, solid, lustrous greenish-black, with a narrow edging of silvery white, uniform in width, extending around point of each feather; shafts, black; plumage in front of hackle, black.

WINGS: Bows, silvery white; coverts, lustrous greenish-black, forming a well defined bar of this color across wing when folded; primaries, black, except a narrow edging of white on lower edge of lower webs; secondaries, black, except lower half of lower webs, which should be white, except near ends of feathers, at which point the white terminates abruptly, leaving ends of feathers black.

BACK: Silvery white, free from brown; saddle, silvery white, with a black stripe in each feather, tapering to a point near its lower extremity.

TAIL: Black; sickles and coverts, lustrous greenish-black; smaller coverts, lustrous greenish-black, edged with white.

BREAST: Black.

BODY AND FLUFF: Body, black; fluff, black, slightly tinged with gray.

LEGS AND TOES: Thighs, black; shank feathers, black; shanks and toes, yellow; toe feathers, black.

UNDER-COLOR OF ALL SECTIONS: Slate.

COLOR OF FEMALE.

HEAD: Plumage, silvery gray.

BEAK: Dark horn shading to yellow at point.

EYES: Reddish-bay.

COMB, FACE, WATTLES AND EAR-LOBES: Bright red.

NECK: Silvery white, center portion of feathers black, slightly penciled with gray; feathers in front of neck, same as breast.

WINGS: Shoulders, bows and coverts, gray, with distinct dark pencilings, outlines of which conform to shape of feathers; primaries, black, with narrow edging of gray penciling on lower webs; secondaries, upper webs black, lower webs gray, with distinct dark pencilings extending around outer edge of feathers.

BACK: Gray, with distinct dark pencilings, outlines of which conform to shape of feathers; feathers, free from white shafts.

TAIL: Black, except the two top feathers, which are penciled on upper edge; coverts, gray, with distinct dark pencilings, outlines of which conform to shape of feathers.

BREAST: Gray, with distinct dark pencilings, outlines of which conform to shape of feathers.

BODY AND FLUFF: Body, gray, with distinct dark pencilings, reaching well down on thighs; fluff, gray, penciled with a darker shade.

LEGS AND TOES: Thighs, gray, with distinct pencilings; shank and toe feathers, same shade of gray; toe feathers, penciled; shanks and toes, dusky yellow.

UNDER-COLOR OF ALL SECTIONS: Slate.

Note: Each feather in back, breast, body, wing-bows and thighs to have three or more distinct pencilings.

DARK BRAHMA MALE

DARK BRAHMA FEMALE

COCHINS.

The Cochin male is very deep bodied, massive in appearance, and maintains characteristic dignity of mien. The breast is carried low while the saddle or cushion is held well up and the whole body shows a tendency to lean forward. The outlines of every section should be well rounded and free from flat or concave surfaces. The extraordinary profusion of long, loose plumage, and the great abundance of downy fiber in the under fluff, produce a large, bulky appearance, conveying the idea of even greater weight than exists. Hard or close-fitting plumage is a very serious defect. The Cochin female should correspond, in a feminine way, with the male, but should be shorter and rounder, possessing a more plump appearance. The back of the Cochin female seems shorter than it really is, owing to the difference in character of the neck and cushion plumage. The cushion is more pronounced than in the male, being very large and full, rising from the cape, and very nearly covering the tail. The head, comb, wattles, and ear-lobes are much smaller than in the male, and are of fine texture.

Cochin Disqualifications.

Vulture hocks, plucked hocks, bare middle toes.

STANDARD WEIGHTS.

Cock ....11 lbs.   Hen .... 8½ lbs.
Cockerel .... 9 lbs.   Pullet .... 7 lbs.

SHAPE OF MALE.

HEAD: Rather short, broad, deep, prominent over eyes, juncture with neck well defined; face of fine texture.

BEAK: Short, stout at base, curving to point.

EYES: Medium large, mild in expression.

COMB: Single, of medium size, set firmly on head, stout at base, upright, straight from front to rear, low in front; serrations moderately deep, dividing comb into five regular points, the middle one highest; free from wrinkles; fine in texture.

WATTLES AND EAR-LOBES: Wattles, rather long, well rounded at lower edges, thin and fine in texture. Ear-lobes, smooth, well defined, hanging about two-thirds as long as wattles.

NECK: Short, full, well proportioned, beautifully arched from rear of head to back; hackle, very long and abundant, flowing well over shoulder and cape.

WINGS: Small, carried well up and well folded; fronts, embedded in plumage of breast; tips, nicely tucked under saddle plumage; wing-bows, smooth and exceedingly well rounded; loose or hanging flights very objectionable.

BACK: Short in appearance, very broad and well rounded; shoulders, very broad, flat under hackle; saddle or cushion rising at base of hackle or cape, very broad and round; plumage, very profuse and long; saddle feathers, flowing over tips of wings and mingling with fluff and under-plumage of tail.

TAIL: Main tail, short, well spread at base, filled underneath with profusion of soft feathers; sickles, short, well rounded and enveloped by coverts and smaller sickles, showing as little of the stiff feathers as possible; saddle and tail to have soft, round, bulky appearance.

BREAST: Carried forward, very full, well rounded, of great breadth and depth.

BODY AND FLUFF: Body, of moderate length, broad, deep, well rounded from point of breast to abdomen, well let down between the legs, broad and well rounded from breast bone to tail, depending more on length of feathers for fullness than on muscular development; fluff, full, soft, abundant.

LEGS AND TOES: Thighs, of moderate length; very strong, large, straight, set well apart; the more long, soft, outstanding plumage, extending well down the shanks and covering knee or hock joints, the better, having the appearance of two great globes of feathers concealing the legs from view; hocks, covered with flexible feathers, curving inward about the joints; free from vulture-like feathering; shanks, short, stout, in bone; plumage, long, beginning just below hocks and covering front and outer sides of shanks, from which it should be outstanding, the upper part growing out from under thigh plumage and continuing into foot feathering. There should be no marked break in the outlines between the plumage of these sections; they should merge naturally into each other and blend together; toes, straight, stout, well spread; middle and outer toes, heavily feathered to ends.

SHAPE OF FEMALE.

HEAD: Neat looking, fairly full in skull, fashioned after that of male, except finer in form; face of fine texture.

BEAK: Short, stout at base, curving to point.

EYES: Of medium size, mild in expression.

COMB: Single, small, straight and upright; nicely rounded to conform to shape of head; divided into five points; free from wrinkles, fine in texture.

WATTLES AND EAR-LOBES: Wattles, small, nicely rounded; fine in texture. Ear-lobes, oblong, fairly well defined, fine in texture.

NECK: Short, nicely arched; plumage, very full, flowing well over shoulders and cape.

WINGS: Small, well folded; fronts, embedded in plumage of breast; tips, concealed between cushion and thigh plumage; wing-bows, smooth and exceedingly well rounded; loose or hanging flights very objectionable.

BACK: Short in appearance, very broad, well rounded; shoulders, broad, flat under neck feathers; cushion, rising from cape, large, full and round; plumage, profuse, flowing over tips of wings well into thigh plumage and almost covering tail feathers.

TAIL: Short, broad at base, carried rather low; well filled underneath with profusion of soft feathers and nearly enveloped by tail coverts, which help to form the cushion.

BREAST: Carried low in front, full, well rounded, of great breadth and depth.

BODY AND FLUFF: Body, of medium length, broad, deep, full and well rounded from point of breast to abdomen, well let down between legs, full and round from breast bone to tail, with great length and fullness of feather; fluff, full, soft, profuse.

LEGS AND TOES: Thighs, of medium length, moderately large, straight, strong, set well apart; with great profusion of long, soft, outstanding fluff plumage, completely hiding hocks or knee-joints, and covering shanks almost to feet; hocks should be well covered with profusion of soft, flexible feathers, curving inward about knee-joints; free from vulture-like feathers; shanks, short, stout in bone, covered profusely with long plumage; toes, straight, well spread; middle and outer toes, heavily feathered to ends.

BUFF COCHINS.

Disqualifications.

Shanks other than yellow. (See general and Cochin disqualifications.)

COLOR OF MALE.

BEAK: Yellow.

EYES: Reddish-bay.

COMB, FACE, WATTLES AND EAR-LOBES: Bright red.

SHANKS AND TOES: Rich yellow.

PLUMAGE: Surface throughout an even shade of rich, golden buff, free from shafting or mealy appearance; the head, neck, hackle, back, wing-bows, and saddle, richly glossed; under-color, a lighter shade free from foreign color. Different shades of buff in two or more sections is a serious defect. A harmonious blending of buff in all sections is most desirable.

COLOR OF FEMALE.

BEAK: Yellow.

EYES: Reddish-bay.

COMB, FACE, WATTLES AND EAR-LOBES: Bright red.

SHANKS AND TOES: Rich yellow.

PLUMAGE: Surface throughout an even shade of rich, golden buff, free from shafting or mealy appearance; the head and neck plumage showing luster of the same shade as the rest of the plumage; under-color a lighter shade, free from foreign color. Different shades of buff in two or more sections is a serious defect. A harmonious blending of buff in all sections is most desirable.

PARTRIDGE COCHINS.

Disqualifications.

Positive white in main tail feathers, sickles or secondaries; shanks other than yellow or dusky yellow. (See general and Cochin disqualifications.)

COLOR OF MALE.

HEAD: Plumage, bright red.

BEAK: Dark horn shading to yellow at point.

EYES: Reddish-bay.

BUFF COCHIN MALE

BUFF COCHIN FEMALE

COMB, FACE, WATTLES AND EAR-LOBES: Bright red.

NECK: Hackle, web of feathers, solid, lustrous greenish-black with a narrow edging of rich, brilliant red, uniform in width, extending around point of each feather; shafts, black; plumage in front of hackle, black.

WINGS: Bows, rich brilliant red; fronts, black; coverts, lustrous greenish-black, forming a well defined bar of this color across wing when folded; primaries, black, lower edges, reddish-bay; secondaries, black, outside web reddish-bay, terminating with greenish-black at end of each feather.

BACK AND SADDLE: Rich, brilliant red with lustrous greenish-black stripe down the middle of each feather, same as in hackle.

TAIL: Black; sickles and smaller sickles, lustrous greenish-black; coverts, lustrous greenish-black edged with rich brilliant red.

BREAST: Lustrous black.

BODY AND FLUFF: Body, black; fluff, black slightly tinged with red.

LEGS AND TOES: Thighs, black; shank and toe feathers, black; shanks and toes, yellow.

UNDER-COLOR OF ALL SECTIONS: Slate.

COLOR OF FEMALE.

HEAD: Plumage, mahogany-brown.

BEAK: Dark horn shading into yellow at point.

EYES: Reddish-bay.

COMB, FACE, WATTLES AND EAR-LOBES: Bright red.

NECK: Reddish-bay; center portion of feathers, black slightly penciled with mahogany-brown; feathers in front of neck, same as breast.

WINGS: Bows and coverts, mahogany-brown penciled with black; primaries, black lacing with edging of mahogany-brown on outer webs; secondaries, inner webs black, outer webs, mahogany-brown penciled with black, the outlines of pencilings conforming to shape of feathers.

BACK: Mahogany-brown distinctly penciled with black, the outlines of pencilings conforming to shape of feathers.

TAIL: Black, the two top feathers, black, penciled with mahogany-brown on upper edges; coverts, mahogany-brown penciled with black.

BREAST: Mahogany-brown distinctly penciled with black, the outlines of pencilings conforming to shape of feathers.

BODY AND FLUFF: Body, mahogany-brown penciled with black; fluff, mahogany-brown.

LEGS AND TOES: Thighs, mahogany-brown penciled with black; shank and toe feathers, mahogany-brown penciled with black; shanks and toes, yellow or dusky yellow.

UNDER-COLOR OF ALL SECTIONS: Slate.

Note: Each feather in back, breast, body, wing-bows, coverts, and thighs to have three or more distinct pencilings.

WHITE COCHINS.

Disqualifications.

Feathers other than white in any part of plumage; shanks other than yellow. (See general and Cochin disqualifications.)

COLOR OF MALE AND FEMALE.

BEAK: Yellow.

EYES: Reddish-bay.

COMB, FACE, WATTLES AND EAR-LOBES: Bright red.

SHANKS AND TOES: Yellow.

PLUMAGE: Web, fluff and quills of feathers in all sections, pure white.

BLACK COCHINS.

Disqualifications.

Feathers other than black in any part of plumage, except in foot or toe feathering; shanks other than yellow or black gradually shading into yellow; bottoms of feet other than yellow. (See general and Cochin disqualifications.)

COLOR OF MALE AND FEMALE.

BEAK: Black shaded with yellow.

EYES: Reddish-bay.

COMB, FACE, WATTLES AND EAR-LOBES: Bright red.

SHANKS AND TOES: Yellow or black gradually shading into yellow; bottoms of feet, yellow.

PLUMAGE: Surface a lustrous greenish-black throughout.

UNDER-COLOR OF ALL SECTIONS: Dull black.

129

PARTRIDGE COCHIN MALE

PARTRIDGE COCHIN FEMALE

131

LANGSHANS.

The general characteristics of the Langshans are great proportionate depth of body, round contour of breast and fineness of bone for size of fowl. The male develops great length of tail feathers, the sickles not uncommonly attaining a length of sixteen or seventeen inches. Its large, well spread tail, carried erect with abundant, close-lying saddle feathers, full-hackled neck and upright carriage, give the effect of a short back. The surface of plumage throughout is close and smooth, being very brilliant with greenish reflections in the black and pure white in the white variety. The body, in both sexes of both varieties, should be evenly balanced on firm, straight legs, with very little backward bend at the knee-joints. The legs are sinewy, the toes long and slender, free from coarseness and the middle toes should be devoid of feathers. The height of the Langshans should be gained by depth of body and erectness of carriage and not from what may be described as stiltiness of legs. Close-standing knee-joints, and narrowness of body are highly objectionable.

Langshan Disqualifications.

Yellow skin, bottom of feet yellow in any part.

STANDARD WEIGHTS.

Cock ....9½ lbs. Hen ....7½ lbs.
Cockerel ....8 lbs. Pullet ....6½ lbs.

SHAPE OF MALE.

HEAD: Of medium size, rather broad.

BEAK: Stout at base, well curved.

EYES: Moderately large, round.

COMB: Single; of medium size, straight, upright, evenly serrated, having five points, not conforming closely to neck, fine in texture.

WATTLES AND EAR-LOBES: Wattles, of moderate length, well rounded. Ear-lobes, oblong, well developed.

NECK: Of good length, well arched; hackle, abundant, flowing well over shoulders.

WINGS: Of medium size, well folded, carried closely to body.

BACK: Short, broad, flat at shoulders, rising from middle of back in a decidedly sharp concave sweep to tail; saddle feathers, abundant, flowing over sides.

TAIL: Long, large, full, well spread at base, carried at an angle of seventy degrees above the horizontal (see illustration, fig. 39); sickles, long, extending decidedly beyond the tail; coverts, long—the longer the better.

BREAST: Broad, round, deep.

BODY AND FLUFF: Body, rather broad and deep in front of thighs; fluff, fairly developed, but not so abundant as to hide profile of hocks.

LEGS AND TOES: Thighs, rather long, strong, well covered with soft feathers; shanks, rather long, stout in bone, straight, set well apart, feathered down the outer sides; toes, long, straight, slender; outer toes, feathered to the end; middle toes, free from feathers.

SHAPE OF FEMALE.

HEAD: Similar to that of male, but smaller.

BEAK: Stout at base, well curved.

EYES: Moderately large, round.

COMB: Single; smaller than that of male, straight and upright, evenly serrated, having five points, fine in texture.

WATTLES AND EAR-LOBES: Wattles, fairly developed, well rounded; fine in texture. Ear-lobes, fairly developed, oblong, fine in texture.

NECK: Of good length, full feathered.

WINGS: Of good size, well folded, carried close to body.

BACK: Of medium length, broad, flat at shoulders, rising from middle of back in a sharp, concave sweep ending well up on tail.

TAIL: Rather long, well spread at base, carried at an angle of sixty-five degrees above the horizontal (see illustration, fig. 39); carried well above and beyond the cushion and furnished with long coverts.

BREAST: Broad, round, deep.

BODY AND FLUFF: Body, rather short, broad, deep, well balanced; fluff, well developed.

LEGS AND TOES: Thighs, rather long, strong, well covered with soft feathers; shanks, of good length, small-boned, set well apart, feathered down outer sides; toes, long, straight, slender; outer toes, feathered to ends; middle toes, free from feathers.

133

BLACK LANGSHAN MALE

134

BLACK LANGSHAN FEMALE

BLACK LANGSHANS.

Disqualifications.

One-half inch or more of white in any part of the plumage, except in foot or toe feathering. (See general and Langshan disqualifications.)

COLOR OF MALE AND FEMALE.

BEAK: Dark horn shading to pinkish tint near lower edge.
EYES: Black or dark brown.
COMB, FACE, WATTLES AND EAR-LOBES: Bright red.
SKIN ON BODY: Pinkish-white.
SHANKS AND TOES: Bluish-black, showing pink between scales; web and bottoms of feet, pinkish-white.
PLUMAGE: Surface, lustrous greenish-black.
UNDER-COLOR: Black.

WHITE LANGSHANS.

Disqualifications.

Feathers other than white in any part of plumage. (See general and Langshan disqualifications.)

COLOR OF MALE AND FEMALE.

BEAK: Light slate-blue shading to pinkish-white.
EYES: Dark brown.
COMB, FACE, WATTLES AND EAR-LOBES: Bright red.
SHANKS AND TOES: Slaty-blue, showing pink between the scales.
PLUMAGE: Web, fluff and quills of feathers in all sections, pure white.

MEDITERRANEAN.

Breeds		*Varieties*
LEGHORNS.............................	{	Single Comb Brown
		Rose Comb Brown
		Single Comb White
		Rose Comb White
		Single Comb Buff
		Rose Comb Buff
		Single Comb Black
		Silver
		Red Pyle
MINORCAS.............................	{	Single Comb Black
		Rose Comb Black
		Single Comb White
		Rose Comb White
		Single Comb Buff
SPANISH.............................		White-Faced Black
BLUE ANDALUSIANS		
ANCONAS.............................	{	Single Comb
		Rose Comb

SCALE OF POINTS, MEDITERRANEAN CLASS.

Symmetry	4
Weight	4
Condition	4
Comb	10
Head—Shape 2, Color 4..................................	6
Eyes—Shape 2, Color 2..................................	4
Beak—Shape 2, Color 2..................................	4
Wattles and Ear-Lobes—Shape 4, Color 6..................	10
Neck—Shape 3, Color 5..................................	8
Wings—Shape 4, Color 6.................................	10
Back—Shape 5, Color 5..................................	10
Tail—Shape 5, Color 4....	9
Breast—Shape 4, Color 4................................	8
Body and Fluff—Shape 3, Color 2........................	5
Legs and Toes—Shape 2, Color 2.........................	4
	100

LEGHORNS.

This breed originated in Italy. It comprises a group characterized by rather small size, yellow legs, white ear-lobes and great activity. Leghorns are hardy and prolific. The males are very alert in carriage. The females are non-sitters, very few of them showing a tendency to broodiness. The leghorn is essentially a breed of alert carriage and graceful curves, and should be bred strictly on these lines. Short backs, short shanks and short bodies are objectionable. The nine varieties of Leghorns should be identical, except in color and comb, while the wide range of plumage enables all admirers of the Leghorn to indulge their fancy. Each variety has its points of beauty in plumage, some of which are exceedingly difficult to produce, thus presenting intricate problems for the breeders to master.

Leghorn Disqualifications.

Red covering more than one-third the surface of ear-lobes in cockerels and pullets. (See general disqualifications.)

STANDARD WEIGHTS.

Cock ....5½ lbs. Hen 4   lbs.
Cockerel ....4½ lbs. Pullet ....3½ lbs.

SHAPE OF MALE.

HEAD: Moderate in length, fairly deep. Face; smooth, fine in texture.

BEAK: Not too long, nicely curved.

EYES: Of medium size and nearly round.

COMB: Single; of medium size, straight and upright, firm and even on head, having five distinct points, deeply serrated and extending well over back of head with no tendency to follow shape of neck; smooth, free from twists, folds and excrescences.

COMB: Rose; of medium size, square in front, firm and even on head, tapering evenly from front to rear and terminating in a well-developed spike, which extends horizontally well back of head; flat, free from hollow center and covered with small, rounded points.

WATTLES AND EAR-LOBES: Wattles, long, thin, even in length, well rounded, smooth in texture, free from folds or wrinkles.

138

Ear-lobes, oval in shape but rather broad, smooth, of moderate size, fitting closely to head.

NECK: Long, nicely arched, hackle abundant, flowing well over the shoulders.

WINGS: Large, well folded.

BACK: Rather long, slightly rounded at shoulders, slightly sloping downward from shoulders to center of back, then rising in a gradually increasing concave sweep to tail.

TAIL: Large, well spread; main tail feathers, carried at an angle of forty-five degrees above the horizontal. (See illustration, figures 39 and 40.)

BREAST: Well rounded, carried well forward.

BODY AND FLUFF: Body, of moderate length and fairly deep; carried nearly horizontal but sloping very slightly from front to rear; fluff, short.

LEGS AND TOES: Thighs and shanks. moderately long, slender; toes, medium length, straight, rather slender.

SHAPE OF FEMALE.

HEAD: Moderate in length and fairly deep; face nearly round, smooth, fine in texture.

BEAK: Not too long, nicely curved.

EYES: Of medium size and nearly round.

COMB: Single, medium in size; deeply serrated, having five distinct points, the front portion of the comb and first point to stand erect and the remainder of comb drooping gradually to one side; fine in texture; free from folds or wrinkles.

COMB: Rose, small, square in front, firm and even on head, tapering evenly from front to rear and terminating in a well developed spike, which extends horizontally back of head; flat, free from hollow center, covered with very small, rounded points.

WATTLES AND EAR-LOBES: Wattles, of moderate size, of even length, thin, well-rounded. Ear-lobes, oval in shape, smooth, thin, free from folds or wrinkles, fitting closely to head.

NECK: Long, slender, gracefully arched.

WINGS: Large, well folded.

BACK: Rather long, slightly rounded, with a slight slope downward from shoulders to center of back, and then rising in a concave incline to tail.

TAIL: Long, full, well spread, carried at an angle of forty degrees above the horizontal. (See illustration, figure 39.)

BREAST: Well rounded, carried well forward.

SINGLE COMB BROWN LEGHORN MALE

SINGLE COMB BROWN LEGHORN FEMALE

BODY AND FLUFF: Body, moderately long, fairly deep, carried nearly horizontal but sloping very slightly downward from front to rear; fluff, rather short.

LEGS AND TOES: Thighs and shanks, moderately long, slender; toes, medium length, straight, slender.

BROWN LEGHORNS.

(Single and Rose Combs.)
Disqualifications.

Positive white in main tail feathers, sickles or secondaries; shanks other than yellow. (See general and Leghorn disqualifications.)

COLOR OF MALE.

HEAD: Plumage, dark red.

BEAK: Horn.

EYES: Reddish-bay.

COMB, FACE AND WATTLES: Bright red.

EAR-LOBES: White or creamy-white.

NECK: Rich, brilliant red, with a lustrous greenish-black stripe running nearly parallel with edges and extending through the middle of each feather, tapering to a point near its lower extremity; plumage in front of hackle, lustrous greenish-black.

WINGS: Bows, rich, brilliant red; fronts, black; coverts, lustrous greenish-black, forming well defined wing-bar when wing is folded; primaries, black, lower webs edged with brown; secondaries, black, edges of lower webs a rich brown of sufficient width to secure wing-bay of same color.

BACK: Rich, brilliant red; saddle feathers, rich, brilliant red with a lustrous greenish-black stripe running through the middle of each feather, same as in hackle.

TAIL: Main tail feathers, black; sickles and coverts, lustrous greenish-black.

BREAST: Lustrous black.

BODY AND FLUFF: Black.

LEGS AND TOES: Thighs, black; shanks, yellow; toes, yellow or dusky yellow.

UNDER-COLOR OF ALL SECTIONS: Slate.

COLOR OF FEMALE.

HEAD: Plumage, golden yellow tinged with dark brown.

BEAK: Horn.

EYES: Reddish-bay.

142

COMB, FACE AND WATTLES: Bright red.

EAR-LOBES: White or creamy-white.

NECK: Golden yellow, with black stripe extending down middle of each feather and tapering to a point near its lower extremity; feathers in front of neck, rich salmon.

WINGS: Bows and coverts, same color as described for back; primaries, slaty-brown, the outer webs slightly edged with brown; secondaries, brown, the outer webs finely stippled with lighter brown.

BACK: Web of feathers, on surface, light brown finely stippled with darker brown, the lighter shade predominating; more importance is attached to fineness, sharp definition of stippling, evenness of color and freedom from shafting than to the particular shade of color, but it is important that the effect produced be that of a soft, even brown that is not suggestive of gray as one extreme to be avoided, or of red, as the other; the unexposed portion of the feather to be brown shading into slate.

TAIL: Dull black, except the two highest main tail feathers which are stippled with lighter brown; coverts, same as back.

BREAST: Rich salmon, shading off lighter under body; free from shafting.

BODY AND FLUFF: Body, light brown stippled with a darker brown, free from shafting; fluff, slate plentifully tinged with brown.

LEGS AND TOES: Thighs, slate plentifully tinged with brown; shanks, yellow; toes, yellow or dusky yellow.

UNDER-COLOR OF ALL SECTIONS: Slate.

WHITE LEGHORNS.

(Single and Rose Combs.)

Disqualifications.

Feathers other than white in any part of plumage; shanks other than yellow. (See general and Leghorn disqualifications.)

COLOR OF MALE AND FEMALE.

BEAK: Yellow.

EYES: Reddish-bay.

COMB, FACE AND WATTLES: Bright red.

EAR-LOBES: White.

SHANKS AND TOES: Rich yellow.

PLUMAGE: Web, fluff and quill of feathers in all sections, pure white.

SINGLE COMB WHITE LEGHORN MALE

SINGLE COMB WHITE LEGHORN FEMALE

ROSE COMB WHITE LEGHORN MALE

ROSE COMB WHITE LEGHORN FEMALE

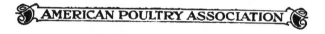

BUFF LEGHORNS.

(Single and Rose Combs.)
Disqualifications.

Shanks other than yellow. (See general and Leghorn disqualifications.)

COLOR OF MALE.

BEAK: Yellow.
EYES: Reddish-bay.
COMB, FACE AND WATTLES: Bright red.
EAR-LOBES: White or creamy-white.
SHANKS AND TOES: Rich yellow.
PLUMAGE: Surface throughout an even shade of rich, golden buff, free from shafting or mealy appearance; head and neck, hackle, back, wing-bow and saddle, richly glossed; under-color, a lighter shade, free from foreign color. Different shades of buff in two or more sections is a serious defect. A harmonious blending of buff in all sections is most desirable.

COLOR OF FEMALE.

BEAK: Yellow.
EYES: Reddish-bay.
COMB, FACE AND WATTLES: Bright red.
EAR-LOBES: White or creamy-white.
SHANKS AND TOES: Rich yellow.
PLUMAGE: Surface throughout an even shade of rich, golden buff, free from shafting or mealy appearance; head and neck plumage showing a luster of the same shade as the rest of the plumage; under-color, a lighter shade, free from foreign color. Different shades of buff in two or more sections is a serious defect. A harmonious blending of buff in all sections is most desirable.

BLACK LEGHORNS.

Disqualifications.

Feathers other than black in any part of plumage; shanks, other than yellow or yellowish-black. (See general and Leghorn disqualifications.)

COLOR OF MALE AND FEMALE.

BEAK: Yellow or dusky yellow.
EYES: Reddish-bay.
COMB, FACE AND WATTLES: Bright red.
EAR-LOBES: White.
SHANKS AND TOES: Yellow or dusky yellow.
PLUMAGE: Surface, lustrous greenish-black; under color of all sections, dull black.

SILVER LEGHORNS.

Disqualifications.

Solid red feathers in any part of plumage; shanks other than yellow or dusky yellow. (See general and Leghorn disqualifications.)

COLOR OF MALE.

HEAD: Plumage, silvery white.
BEAK: Yellow.
EYES: Reddish-bay.
COMB, FACE AND WATTLES: Bright red.
EAR-LOBES: White or creamy-white.
NECK: Hackle, silvery white with narrow black stripe extending down middle of each lower hackle feather, tapering to a point near its lower extremity; plumage in front of hackle, rich glossy black.
WINGS: Fronts, black; shoulders, black; bows, silvery white; coverts, black, forming a distinct bar across the wing; primaries, black, except the lower feathers, the outer edges of which should be silvery white; secondaries, part of outer webs of feathers in wing-bay, white, remainder of feathers, black.

SINGLE COMB BUFF LEGHORN MALE

SINGLE COMB BUFF LEGHORN FEMALE

BACK: Silvery white; cape, black; saddle feathers, silvery white.

TAIL: Black; sickles, lustrous black; upper coverts, lustrous black; lower coverts, silvery white.

BREAST: Black.

BODY AND FLUFF: Black.

LEGS AND TOES: Thighs, black; shanks and toes, yellow.

UNDER-COLOR OF ALL SECTIONS: Gray.

COLOR OF FEMALE.

HEAD: Plumage, silvery gray.

BEAK: Yellow.

EYES: Reddish-bay.

COMB, FACE AND WATTLES: Bright red.

EAR-LOBES: White or creamy-white.

NECK: Silvery gray, with narrow black stripe extending down middle of each feather, tapering to a point near its lower extremity; feathers in front of neck, light salmon.

WINGS: Bows and coverts, gray, formed of silvery white finely stippled with ashy gray, free from dark marks or bars; primaries and secondaries, upper webs, gray, lower webs, slaty-gray.

BACK: Gray, formed of silvery white finely stippled with ashy-gray.

TAIL: Black, except the upper two feathers, which are light gray; coverts, gray.

BREAST: Salmon, shading to gray towards sides.

BODY AND FLUFF: Body, gray, formed of silvery white finely stippled with ashy-gray, free from dark marks or bars; fluff, light ashy-gray.

LEGS AND TOES: Thighs, gray, formed of silvery white finely stippled with ashy-gray; shanks and toes, yellow.

UNDER-COLOR OF ALL SECTIONS: Gray.

RED PYLE LEGHORNS.

Disqualifications.

Shanks other than yellow in cockerel and pullets. (See general and Leghorn disqualifications.)

COLOR OF MALE.

HEAD: Plumage, bright orange.
BEAK: Yellow.
EYES: Reddish-bay.
COMB, FACE AND WATTLES: Bright red.
EAR-LOBES: White or creamy-white.
NECK: Hackle, light orange, which may be tinged with rich yellow; plumage in front of hackle, white.
WINGS: Shoulders, white; fronts, white; bows, red; coverts, white, forming distinct bar across wing; primaries, white, except lower feathers, outer web of which is bay; secondaries, part of outer web forming the wing-bay, red, remainder of feathers, white.
BACK: Red; saddle, light orange.
TAIL: Main tail, sickles and coverts, white.
BREAST: White.
BODY AND FLUFF: White.
LEGS AND TOES: Thighs, white; shanks and toes, yellow.
UNDER-COLOR OF ALL SECTIONS: White.

COLOR OF FEMALE.

HEAD: Plumage, golden.
BEAK: Yellow.
EYES: Reddish-bay.
COMB, FACE AND WATTLES: Bright red.
EAR-LOBES: White or creamy-white.
NECK: White, the feathers edged with gold; feathers in front of neck, white tinged with salmon.
WINGS: White.
BACK: White.
TAIL: White.
BREAST: Salmon.
BODY AND FLUFF: White.
LEGS AND TOES: Thighs, white; shanks and toes, yellow.
UNDER-COLOR OF ALL SECTIONS: White.

153

MINORCAS.

Minorcas are the largest of the Mediterranean breeds. They were originally called Red-Faced Black Spanish. They are distinguished by long bodies, large combs, long wattles and large white ear-lobes. The Minorca back should be long and should slope slightly downward to tail; the tail, large and full and only moderately elevated. Their legs should be firm, muscular and set squarely under the long, powerful looking body. The Minorca plumage is compact, smooth on surface and fits closely to the body. The lustrous greenish-black surface color of the Black variety should be free from purple; the plumage of the White variety should be a pure white, free from foreign color; the Buffs should be one uniform shade of rich, golden-buff throughout.

STANDARD WEIGHTS.

(*Single Comb Black.*)

Cock 9   lbs.   Hen ....7½ lbs.
Cockerel ....7½ lbs.   Pullet ....6½ lbs.

(*Single Comb White, Rose Comb Black, Single Comb Buff, and Rose Comb White.*)

Cock 8   lbs.   Hen ....6½ lbs.
Cockerel ....6½ lbs.   Pullet ....5½ lbs.

SHAPE OF MALE.

HEAD: Moderately long, wide, deep. Face, full and smooth.
BEAK: Of good length, stout, well curved.
EYES: Large, oval.
COMB: Single; large, straight and upright, fine and even on head, smooth, deeply and evenly serrated, having six regular and distinct points, the middle points the longest and same in length as width of blade; front not to extend beyond point half-way between nostrils and point of beak, but extending well over back of head, with a tendency to follow shape of neck.
COMB: Rose; moderately large, square in front, not covering nostrils; firm and even on head, tapering evenly from front to rear, terminating in a well-developed spike, which extends well back of

head and inclines slightly below the horizontal; top, flat, free from hollow center, and covered with small, rounded points.

WATTLES AND EAR-LOBES: Wattles, long, large, thin, free from folds and wrinkles, smooth and fine in texture. Ear-lobes, large, almond-shape, smooth, thin, free from folds and wrinkles, fitting closely to head.

NECK: Rather long, arched; hackle, abundant, flowing well over shoulders.

WINGS: Large, well folded.

BACK: Long, flat at shoulders, broad and rounded at sides, sloping slightly downward to tail; saddle feathers, long.

TAIL: Large and full, main tail feathers carried at an angle of forty degrees above the horizontal (see illustration, figure 39); sickles, large, long, well curved; coverts, abundant.

BREAST: Deep, well rounded and prominent.

BODY AND FLUFF: Body, long, broad, deep, straight from front to rear; fluff, rather short.

LEGS AND TOES: Thighs, of medium length, stout; shanks, rather long, straight and strong, set well apart; toes, straight.

SHAPE OF FEMALE.

HEAD: Moderately long, wide, deep. Face, full and smooth.

BEAK: Of good length, stout, well curved.

EYES: Large, oval.

COMB: Single; large, forming one loop over beak, then drooping down the opposite side of head; deeply and evenly serrated, with six regular and distinct points.

COMB: Rose; moderately large, square in front, not covering nostrils, firm and even on head, tapering evenly from front to rear, terminating in a well-developed spike, which extends back of head and inclines slightly below the horizontal; top, flat, free from hollow center, and covered with small, rounded points.

WATTLES AND EAR-LOBES: Wattles, long, large, thin, free from folds and wrinkles, smooth and fine in texture. Ear-lobes, large, almond-shaped, smooth, thin, free from folds and wrinkles, fitting closely to head.

NECK: Rather long, slightly arched.

WINGS: Large, well folded.

BACK: Long, flat at shoulders, broad, and rounded at sides, sloping slightly downward to tail.

TAIL: Long, full and carried at an angle of forty degrees above the horizontal. (See illustration, figure 39.)

SINGLE COMB BLACK MINORCA MALE

156

SINGLE COMB BLACK MINORCA FEMALE

157

ROSE COMB BLACK MINORCA MALE

ROSE COMB BLACK MINORCA FEMALE

BREAST: Deep, well rounded, prominent.

BODY AND FLUFF: Body, long, broad, deep, straight from front to rear; fluff, rather short.

LEGS AND TOES: Thighs, of medium length, stout; shanks, rather long, straight and strong, well set apart; toes, straight.

BLACK MINORCAS.

(*Single and Rose Combs.*)

Disqualifications.

Red in ear-lobes covering one-third or more of surface; red in any part of plumage; pure white in any part of plumage extending more than one-half inch, or two or more feathers tipped or edged with positive white; shanks, other than dark slate or nearly black. (See general disqualifications.)

COLOR OF MALE AND FEMALE.

BEAK: Black.

EYES: Dark brown.

COMB, FACE AND WATTLES: Bright red.

EAR-LOBES: White.

LEGS AND TOES: Black or dark slate.

PLUMAGE: Surface, lustrous greenish-black throughout.

UNDER-COLOR OF ALL SECTIONS: Dull black.

WHITE MINORCAS.

(*Single and Rose Combs.*)

Disqualifications.

Red in ear-lobes, covering one-third or more of surface; feathers other than white in any part of plumage; shanks other than white or pinkish-white. (See general disqualifications.)

COLOR OF MALE AND FEMALE.

BEAKS: Pinkish-white.

EYES: Reddish-bay.

COMB, FACE AND WATTLES: Bright red.

EAR-LOBES: White.

SHANKS AND TOES: Pinkish-white.

PLUMAGE: Web, fluff and quills of feathers in all sections, pure white.

BUFF MINORCAS

(Single Combs.)

Disqualifications.

Red in ear-lobes covering one-third or more of surface; shanks other than white or pinkish-white. (See general disqualifications.)

COLOR OF MALE.

BEAK: White or light horn.

EYES: Reddish-bay.

COMB, FACE, WATTLES AND EAR-LOBES: Bright red.

EAR-LOBES: White.

SHANKS AND TOES: White or pinkish-white.

PLUMAGE: Surface throughout an even shade of rich, golden buff, free from shafting or mealy appearance; the head, neck, back, wing-bows and saddle, richly glossed. Under-color in all sections, a lighter shade of buff, free from foreign color. Different shades of buff in two or more sections is a serious defect. A harmonious blending of buff in all sections is most desirable.

COLOR OF FEMALE.

BEAK: White or light horn.

EYES: Reddish-bay.

COMB, FACE AND WATTLES: Bright red.

EAR-LOBES: White.

SHANKS AND TOES: White or pinkish-white.

PLUMAGE: Surface throughout an even shade of rich, golden buff, free from shafting or mealy appearance; the head and neck plumage showing a luster of same shade as the rest of plumage. Under-color in all sections, a lighter shade, free from foreign color. Different shades of buff in two or more sections is a serious defect. A harmonious blending of buff in all sections is most desirable.

161

WHITE-FACED BLACK SPANISH MALE

WHITE-FACED BLACK SPANISH FEMALE

163

WHITE-FACED BLACK SPANISH.

The White-Faced Black Spanish is probably the oldest pure-bred fowl in the Mediterranean class. The development of the long, pendulous white face, free from wrinkles has been the aim of breeders of this valuable and attractive breed. In general appearance the Black Spanish are graceful and stylish. The rich, glossy black plumage, the rather large, five-pointed comb and the long pendulous white face give this early claimant to popular favor a distinct individuality among standard-bred fowls.

Disqualifications.

Decided red in the face, except directly above the eyes; face so puffy as to obstruct the sight; foreign color in any part of plumage; shanks other than blue or dark leaden-blue. (See general disqualifications.)

STANDARD WEIGHTS.

Cock ....8 lbs. Hen ....6½ lbs.
Cockerel ....6½ lbs. Pullett ....5½ lbs.

SHAPE OF MALE.

HEAD: Long, broad, deep.

FACE: Long, deep, smooth, free from wrinkles, rising well over eyes in an arched form and not obstructing sight, extending towards back of head and to base of beak, covering cheeks and joining the wattles and ear-lobes; the greater the extent of surface the better.

BEAK: Rather long, stout.

EYES: Large, oval.

COMB: Single; of medium size, straight and upright, firm and even on head, rising from the base of beak and extending in an arch form beyond back of head; deeply and evenly serrated, having five points; very fine in texture.

WATTLES AND EAR-LOBES: Wattles, smooth, very long, thin,

164

ribbon-like. Ear-lobes, very large, free from folds and wrinkles, meeting in front, extending well backward on each side of neck, hanging very low and regularly rounded on lower edges; very smooth.

NECK: Long, well arched; hackle, abundant, flowing well over shoulders.

WINGS: Large, well folded.

BACK: Long, broad and straight, sloping downward to saddle, which rises in a short concave sweep to tail.

TAIL: Large and full; main tail feathers carried at an angle of forty-five degrees above the horizontal; (see illustration, figures 39 and 40); sickles, large, long, well curved; coverts, abundant.

BREAST: Deep, well rounded.

BODY AND FLUFF: Body, long, moderately wide, straight from front to rear; fluff, short.

LEGS AND TOES: Thighs, moderate size, long; shanks, long; toes, straight.

SHAPE OF FEMALE.

HEAD: Long, broad, deep.

FACE: Long, deep, smooth, free from wrinkles, rising well over eyes in an arched form and not obstructing the sight, extending towards back of head and to base of beak, covering the cheeks and joining wattles and ear-lobes; the greater the extent of surface the better.

BEAK: Rather long, stout.

EYES: Large, oval.

COMB: Single, moderately large, deeply serrated, having five points, drooping to one side; very fine in texture.

WATTLES AND EAR-LOBES: Wattles, smooth, very long, thin, ribbon-like. Ear-lobes, very large, free from folds and wrinkles, meeting in front, extending well backward on each side of neck, hanging very low and regularly rounded on lower edges; very smooth.

NECK: Long, well arched.

WINGS: Large, well folded.

BACK: Long, broad, and straight, sloping downward to middle, then rising in a short, concave sweep to tail.

TAIL: Large, carried at an angle of forty-five degrees above the horizontal (see illustration, figure 39); the two main tail feathers slightly curved, especially in pullets.

BREAST: Deep, well rounded.

BODY AND FLUFF: Body, long, moderately wide, straight from front to rear; fluff, short.

LEGS AND TOES: Thighs, moderately long; shanks, long; toes, straight.

COLOR OF MALE AND FEMALE.

BEAK: Black.

EYES: Dark brown.

COMB: Bright red.

FACE: Pure white.

WATTLES AND EAR-LOBES: Wattles; males, bright red, except inside of upper part which is white; females, bright red. Ear-lobes, pure white.

SHANKS AND TOES: Dark leaden-blue or black.

PLUMAGE: Surface, lustrous greenish-black throughout.

UNDER-COLOR OF ALL SECTIONS: Dark slate.

BLUE ANDALUSIANS.

Andalusians are, according to the best available records, natives of Andalusia, a province in Southern Spain. Late in the eighteenth century they were imported to Southwestern England, and to America about 1870. Their type is distinctive. The typical Andalusian should be very symmetrical, graceful and compact, of medium size, fine structure and stately carriage. The ground color of all sections, excepting hackle, back, wing-bows, sickles and tail coverts of males, should be a rich, even, medium shade of slaty-blue which should be defined by a clear and distinct lacing of a darker blue. The neck, back, saddle, wing-bows, sickles and tail-coverts of males should be a very dark lustrous blue. Andalusians that approach the standard requirements are especially attractive.

Disqualifications.

Red in ear-lobes covering one-third or more of surface; red or positive white in color of plumage; shanks other than blue or slaty-blue. (See general disqualifications.)

STANDARD WEIGHTS.

Cock	6 lbs.	Hen	5 lbs.
Cockerel	5 lbs.	Pullet	4 lbs.

SHAPE OF MALE.

HEAD: Moderately long and deep; face, full and smooth, fine in texture.

BEAK: Of moderate length, nicely curved.

EYES: Large, oval.

COMB: Single; of medium size, smooth, straight and upright, firm and even on the head; evenly and deeply serrated, having five points, blade following slightly the curve of neck.

WATTLES AND EAR-LOBES: Wattles, long, thin, smooth. Ear-lobes, almond-shaped, of moderate size, smooth.

NECK: Rather long, well arched, with abundant hackle flowing well over shoulders.

WINGS: Large, well folded.

BACK: Of moderate length, rather broad and high at shoulders, sloping downward to saddle which rises with a concave sweep to tail; saddle feathers, long and abundant.

BLUE ANDALUSIAN MALE

BLUE ANDALUSIAN FEMALE

169

TAIL: Large, full and well spread, main tail feathers carried at an angle of forty degrees above the horizontal (see illustration, figure 39); sickles, long, even, well curved; coverts, abundant.

BREAST: Broad, deep, well rounded, carried well up and forward.

BODY AND FLUFF: Body, deep, well rounded, straight from front to rear; fluff, short.

LEGS AND TOES: Thighs of moderate size, rather long; hock joints showing well below body line; shanks, long, standing well apart; toes, straight.

SHAPE OF FEMALE.

HEAD: Moderately long, deep, full and smooth.

BEAK: Of moderate length, nicely curved.

EYES: Large, oval.

COMB: Single; medium in size, deeply serrated, having five distinct points, the front portion of comb and first point to stand erect and the remainder of comb drooping gradually to one side; fine in texture, free from folds or wrinkles.

WATTLES AND EAR-LOBES: Wattles, moderately long, thin, well rounded. Ear-lobes, almond-shape, of moderate size, smooth.

NECK: Long, gracefully arched.

WINGS: Large, well folded.

BACK: Of moderate length, rather broad and slightly elevated at shoulders, carried on nearly a straight line to rear of cushion, then rising with a rather short curve to tail.

TAIL: Long, well spread, carried at an angle of forty degrees above the horizontal. (See illustration, figure 39.)

BREAST: Broad, deep and well rounded, carried well up and forward.

BODY AND FLUFF: Body, deep, well rounded, straight from front to rear; fluff, short.

LEGS AND TOES: Thighs, of medium size, rather long; shanks, long; toes, straight.

COLOR OF MALE.

HEAD: Plumage, dark slaty-blue.

BEAK: Horn.

EYES: Reddish-bay.

COMB, FACE AND WATTLES: Bright red.

EAR-LOBES: White.

NECK: Very dark lustrous blue.

WINGS: Bows, very dark lustrous blue; coverts, a clear and

even medium shade of slaty-blue, having a well defined lacing of a darker blue; primaries, a clear, even medium shade of slaty-blue; secondaries, inner webs a clear even shade of slaty-blue, outer webs slaty-blue, each feather having a clear and well defined lacing of a darker blue.

BACK: Very dark lustrous blue.

TAIL: Sickles, very dark lustrous blue; tail coverts, lustrous blue; main tail feathers, a clear, even slaty-blue, each feather having a well defined lacing of darker blue.

BREAST: A clear, even, medium shade of slaty-blue, each feather having a clear and well defined lacing of a darker blue.

BODY AND FLUFF: Body, a clear, even, medium shade of slaty-blue, each feather having a clear and well defined lacing of a darker blue; fluff, slaty-blue.

LEGS AND TOES: Thighs, a clear even shade of slaty-blue, each feather having a well defined lacing of a darker blue; shanks and toes, leaden-blue.

UNDER COLOR OF ALL SECTIONS: Slaty-blue.

COLOR OF FEMALE.

HEAD: Plumage, slaty-blue.

BEAK: Horn.

EYES: Reddish-bay.

COMB, FACE AND WATTLES: Bright red.

EAR-LOBES: White.

NECK: Slaty-blue, laced with a darker blue.

WINGS: Primaries, a clear, even medium shade of slaty-blue; remainder of wing an even shade of slaty-blue, darker than that of primaries; feathers in all sections, except primaries, having clear and well defined lacings of a darker blue.

BACK: Slaty-blue, each feather having a clear and well defined lacing of a darker blue.

TAIL: Slaty-blue, each feather laced with a darker blue.

BREAST: Slaty-blue, each feather having a clear and well-defined lacing of a darker blue.

BODY AND FLUFF: Body, slaty-blue, each feather having a clear and well defined lacing of a darker blue; fluff, slaty-blue.

LEGS AND TOES: Thighs, slaty-blue, each feather having a clear and well defined lacing of a darker blue; shanks and toes, leaden-blue.

UNDER-COLOR OF ALL SECTIONS: Slaty-blue.

171

ANCONAS.

(Single and Rose Combs.)

Anconas originated in Italy, and take their name from the city of Ancona. It is one of the oldest breeds of the Mediterranean family and was first imported to America about 1890. The breed resemble the Leghorn in type and they are bred with both single and rose combs, every other characteristic being identical. They are bred on one color scheme only, mottled black and white.

Disqualifications.

Red in ear-lobes, covering more than one-half of surface; red in any part of plumage; shanks other than yellow or yellow mottled with black. (See general disqualifications.)

STANDARD WEIGHTS.

Cock	5½ lbs.	Hen	4½ lbs.
Cockerel	4½ lbs.	Pullet	3½ lbs.

SHAPE OF MALE.

HEAD: Moderate in length, fairly deep; face; smooth, fine in texture, nearly round.

BEAK: Not too long, nicely curved.

EYES: Of medium size and nearly round.

COMB: Single; of medium size, straight and upright, firm and even on head; having five distinct points, deeply serrated and extending well over back of head, with no tendency to follow shape of neck; smooth, free from twists, folds and excrescences.

COMB: Rose; moderately small, square in front, firm and even on head, tapering evenly from front to rear and terminating in a well developed spike which extends horizontally well back of head; flat, free from hollow center, and covered with small rounded points.

WATTLES AND EAR-LOBES: Wattles, long, thin, well rounded, smooth in texture, free from folds and wrinkles. Ear-lobes, a broadened almond-shape, of moderate size, smooth, fitting closely to head.

NECK: Long, nicely arched; hackle, abundant, flowing well over shoulders.

WINGS: Large, well folded.

BACK: Of good length, somewhat rounded at shoulders, slightly sloping downward from shoulders to saddle then rising with a concave sweep to tail.

TAIL: Large, well spread; main tail feathers carried at an

angle of forty-five degrees above the horizontal (see illustrations, figures 39 and 40) ; sickles, long, well curved; coverts, abundant.

BREAST: Well rounded, carried well forward.

BODY AND FLUFF: Body, of moderate length, fairly deep, straight from front to rear; fluff, short.

LEGS AND TOES: Thighs and shanks, moderately long and slender; toes, straight.

SHAPE OF FEMALE.

HEAD: Moderate in length, fairly deep; face, smooth, fine in texture, well rounded.

BEAK: Not too long, nicely curved.

EYES: Medium in size and nearly round.

COMB: Single; medium in size, deeply serrated, having five distinct points, the front portion of comb and first point to stand erect, the remainder of comb drooping gradually on one side; fine in texture, free from folds or wrinkles.

COMB: Rose; small, square in front, firm and even on head, tapering evenly from front to rear and terminating in a well developed spike which extends horizontally back of head; flat, free from hollow center, covered with very small rounded points.

WATTLES AND EAR-LOBES: Wattles, of moderate size, thin, well rounded. Ear-lobes, oval in shape, smooth, thin, free from folds and wrinkles, fitting closely to head.

NECK: Long, slender and gracefully arched.

WINGS: Large and well folded.

BACK: Of good length, somewhat rounded, with slight slope downward from shoulders to center of back and then rising in a concave sweep to tail.

TAIL: Long, full and well-spread, carried at an angle of forty degrees above the horizontal. (See illustration, figure 39.)

BREAST: Well rounded, carried well forward.

BODY AND FLUFF: Body, moderately long, fairly deep, straight from front to rear; fluff, rather short, more developed than in male.

LEGS AND TOES: Thighs and shanks, moderately long, slender; toes, straight.

COLOR OF MALE.

HEAD: Plumage, black.

BEAK: Yellow, upper mandible shaded with black.

EYES: Reddish-bay.

COMB, FACE AND WATTLES: Bright red.

EAR-LOBES: White or creamy-white.

SINGLE COMB ANCONA MALE

SINGLE COMB ANCONA FEMALE

NECK: Lustrous greenish-black, about one feather in five tipped with white; plumage, in front of neck, same as breast.

WINGS: Bows, lustrous greenish-black, about one feather in five ending with a white tip; coverts, lustrous greenish-black, many ending with a white tip; primaries, black, tipped with white; secondaries, black, ending with white tips.

BACK: Lustrous greenish-black, one feather in ten tipped with white; saddle, lustrous greenish-black, ending with white tips.

TAIL: Black, feathers ending with a white tip; sickles and coverts lustrous greenish-black ending with white tips

BREAST: Lustrous black, about one feather in five tipped with white.

BODY AND FLUFF: Body, lustrous black, about one feather in five tipped with white; fluff, black, slightly tinged with white.

LEGS AND TOES: Thighs, black, slightly tipped with white; shanks and toes, yellow or yellow mottled with black.

UNDER-COLOR OF ALL SECTIONS: Dark slate.

COLOR OF FEMALE.

HEAD: Plumage, black.

BEAK: Yellow, upper mandible shaded with black.

EYES: Reddish-bay.

COMB, FACE AND WATTLES: Bright red.

EAR-LOBES: White or creamy-white.

NECK: Lustrous black, about one feather in five tipped with white; feathers in front of neck same as breast.

WINGS: Bows, lustrous black, one feather in five ending with a white tip; coverts, lustrous black, about one feather in five ending with a white tip; primaries, black tipped with white; secondaries, black, ending with white tips.

BACK: Lustrous black, about one feather in five ending with a white tip.

TAIL: Black, feathers ending with a white tip; coverts, black, at least one feather in five ending with a white tip.

BREAST: Lustrous black, about one feather in five ending with a white tip.

BODY AND FLUFF: Body, black, about one feather in five ending with a white tip; fluff, black, slightly tinged with white.

LEGS AND TOES: Thighs, black, slightly tipped with white; shanks and toes, yellow or yellow mottled with black.

UNDER-COLOR OF ALL SECTIONS: Dark slate.

176

ENGLISH.

Breeds	*Varieties*

DORKINGS............................ { White
Silver-gray
Colored

REDCAPS..............................

ORPINGTONS........................ { Single Comb Buff
Single Comb Black
Single Comb White
Single Comb Blue

CORNISH............................ { Dark
White
White-Laced Red

SUSSEX............................. { Speckled
Red

SCALE OF POINTS FOR THE ENGLISH CLASS.

Symmetry	4
Weight	4
Condition	4
Comb	8
Head — Shape 2, Color 2	4
Beak — Shape 2, Color 2	4
Eyes — Shape 2, Color 2	4
Wattles and Ear-Lobes—Shape 2, Color 2	4
Neck — Shape 4, Color 4	8
Wings — Shape 4, Color 6	10
Back — Shape 6, Color 4	10
Tail — Shape 6, Color 4	10
Breast — Shape 6, Color 4	10
Body and Fluff — Shape 5, Color 3	8
Legs and Toes — Shape 5, Color 3	8
	100

DORKINGS.

The Dorking is one of the oldest breeds of domestic fowls. The male is large, and his broad, deep, low-set body, which is nearly rectangular in shape, when viewed from the side, combined with its short legs, gives the bird a very compact and solid appearance. The female closely resembles the male except in plumage, though she is not as large in size and is a trifle shorter in legs, and has a lower-set body. The skin and flesh of Dorkings are white. Dorkings differ from most other breeds in having five toes.

SHAPE OF MALE.

HEAD: Rather large.

BEAK: Of medium length, stout, well curved.

EYES: Prominent.

COMB: Silver-Gray and Colored Dorkings, single; rather large, straight and upright, evenly serrated, having six well defined points, the front and rear points shorter than the other four. White Dorkings, rose; square in front, firm and even on head, terminating in a well defined spike; top, comparatively flat and covered with small, rounded points.

WATTLES AND EAR-LOBES: Wattles, rather large, well rounded at lower end. Ear-lobes, of medium size, about one-half the length of wattles.

NECK: Rather short, arched; hackle, full and abundant, flowing well over shoulders, making it appear very broad, tapering to head.

WINGS: Large, well folded against body, not drooping.

BACK: Broad, long, straight, declining to tail; saddle feathers, abundant.

TAIL: Large, full, somewhat expanded, carried at an angle of forty degrees above the horizontal. (See illustration, figure 39.)

BREAST: Broad, deep, full, well rounded, carried forward.

BODY AND FLUFF: Body, long, broad, deep; fluff, smooth in surface, moderately developed.

LEGS AND TOES: Thighs, large, short, well meated, set well apart; shanks, short, stout, round in bone; toes, five on each foot, front and fifth toes moderately long and smooth, fifth toe well

178

separated from the fourth and directly above it, rising on a slight incline from base to point.

SHAPE OF FEMALE.

HEAD: Of medium size.

BEAK: Of medium length, stout, well curved.

EYES: Prominent.

COMB: Silver-Gray and Colored Dorkings, single; similar to that of male, but much smaller and falling over to one side. White Dorkings, rose; similar to that of male, but much smaller.

WATTLES AND EAR-LOBES: Wattles, uniform, rather broad, well rounded. Ear-lobes, medium size.

NECK: Rather short, arched: feathers full and abundant, flowing over shoulders, tapering to head.

WINGS: Large, folded well against body.

BACK: Broad, long, straight, declining slightly to tail.

TAIL: Well developed, main feathers broad, close together, carried at an angle of forty degrees above the horizontal. (See illustration, figure 39.)

BREAST: Broad, deep, well rounded, carried forward.

BODY AND FLUFF: Body, long, broad, deep, low-set; fluff, smooth in surface, moderately developed.

LEGS AND TOES: Thighs, large, short, well meated; shanks short, stout, round in bone; toes, five on each foot, front and fifth toes moderately long, smooth, fifth toe well separated from fourth and directly over it and rising on a slight incline from base to point.

SILVER-GRAY DORKING MALE

SILVER-GRAY DORKING FEMALE

181

WHITE DORKINGS.

Disqualifications.

Feathers other than white in any part of plumage; shanks other than white or flesh color. (See general disqualifications.)

STANDARD WEIGHTS.

Cock	..............7½ lbs.	Hen	..............6 lbs.
Cockerel	..........6½ lbs.	Pullet	............5 lbs.

COLOR OF MALE AND FEMALE.

BEAK: White.
EYES: Reddish-bay.
COMB, FACE, WATTLES AND EAR-LOBES: Red.
SHANKS AND TOES: White.
PLUMAGE: Web, fluff, and quills of feathers in all sections, pure white.

SILVER-GRAY DORKINGS.

Disqualifications.

Shanks other than white or flesh color. (See general disqualifications.)

STANDARD WEIGHTS.

Cock	..............8 lbs.	Hen	..............6½ lbs.
Cockerel	..........7 lbs.	Pullet	............5½ lbs.

COLOR OF MALE.

HEAD: Plumage, silvery white.
BEAK: White streaked with horn.
EYES: Reddish-bay.
COMB, FACE, WATTLES AND EAR-LOBES: Bright red.
NECK: Hackle, pure silvery white; free from straw tinge or rustiness, very narrow stripes of gray in the lower feathers, permissible, but not desirable; plumage in front of hackle, black.
WINGS: Bows, silvery white; coverts, lustrous greenish-black, forming a wide bar across wing; primaries, quills, black, upper webs black, lower webs white; secondaries, quills, black; upper webs black; lower webs white with a black spot at the end of each feather.

BACK AND SADDLE: Silvery white; cape, black.

TAIL: Sickles, greenish-black; a little white at base of main tail of cock is allowable; coverts, rich, lustrous black; smaller coverts, black.

BREAST: Lustrous black.

BODY AND FLUFF: Black.

LEGS AND TOES: Thighs, black; shanks and toes, white.

UNDER-COLOR OF ALL SECTIONS: Slate.

COLOR OF FEMALE.

HEAD: Plumage, silvery white.

BEAK: White streaked with horn.

EYES: Reddish-bay.

COMB, FACE, WATTLES AND EAR-LOBES: Bright red.

NECK: Silvery white, with a fine, black stripe extending down middle of each feather, tapering to a point near its extremity; feathers in front of neck, reddish salmon.

WINGS: Bows, silvery white finely stippled with ashy-gray; coverts, silvery white stippled with ashy-gray; primaries, upper webs dark slate, lower webs slaty-gray; secondaries, upper webs dark slate, lower webs, slaty-gray.

BACK: Gray, formed of silvery white finely stippled with ashy-gray.

TAIL: Black, penciled with gray on outside and dark slate on inside.

BREAST: Salmon-red, shading to ashy-gray at sides.

BODY AND FLUFF: Body, gray finely stippled with ashy-gray, free from dark marks across feathers; under part of body, gray; fluff, ashy-gray.

LEGS AND TOES: Thighs, ashy-gray; shanks and toes, white.

UNDER COLOR OF ALL SECTIONS: Slate.

COLORED DORKINGS.

Disqualifications.

Shanks other than white or flesh color; absence of fifth toe. (See general disqualifications.)

STANDARD WEIGHTS.

Cock	9 lbs.	Hen	7 lbs.
Cockerel	8 lbs.	Pullet	6 lbs.

COLOR OF MALE.

HEAD: Plumage, very light gray.

BEAK: Dark horn.

EYES: Reddish-bay.

COMB, FACE, WATTLES AND EAR-LOBES: Bright red.

NECK: Light straw color, with a wide, black stripe extending down middle of each feather and terminating in a point near its lower extremity; plumage in front of hackle, black.

WINGS: Bows, light straw; coverts, lustrous greenish-black, forming a wide bar across wing; primaries, black or dark slate; secondaries, upper webs black, lower webs white.

BACK: Cape, black and white; saddle feathers, light straw, with a wide black stripe extending down middle of each feather.

TAIL: Black; sickles, greenish-black; coverts, lustrous black.

BREAST: Lustrous black.

BODY AND FLUFF: Black.

LEGS AND TOES: Thighs, black; shanks and toes, white.

UNDER-COLOR OF ALL SECTIONS: Dark slate.

COLOR OF FEMALE.

HEAD: Plumage, black or nearly black.

BEAK: Dark horn.

EYES: Reddish-bay.

COMB, FACE, WATTLES AND EAR-LOBES: Bright red.

NECK: Black with a narrow edging of gray on front of feathers, the gray extending to sides; feathers in front of neck, dark salmon.

WINGS: Bows, dark gray laced with black; coverts, dark gray laced with black; shafts of feathers, brown; primaries, dark brown; secondaries, upper web black, lower web dark gray.

BACK: Lustrous black; shaft of feathers, light bay.

TAIL: Dark brown pencilled with gray on surface and black on inside.

BREAST: Dark salmon edged with black; shafts of feathers, light bay.

BODY AND FLUFF: Body, dark brown or black, slightly mixed with gray; fluff, dark gray or dull black.

LEGS AND TOES: Thighs, dark gray and brown; shanks and toes, white.

UNDER-COLOR OF ALL SECTIONS: Dark slate.

184

REDCAPS.

This is a breed of English origin. The male should be of good size; have a large rose comb; flowing hackle; a straight back of medium length; large, well expanded tail and the full well rounded breast which is characteristic of this breed. The female should be of good size, and have even, well balanced rose comb; a round, well formed breast; long, well shaped back, and deep, long body. The tail should be of good size and well expanded, giving the specimen a well balanced appearance.

Disqualifications.

Solid white ear-lobes; foreign colored feathers except white in primaries; mottled breast in male; shanks other than slate or leaden-blue in color. (See general disqualifications.)

STANDARD WEIGHTS.

Cock ....7½ lbs.   Hen ....6 lbs.
Cockerel ....6 lbs.   Pullet ....5 lbs.

SHAPE OF MALE.

HEAD: Short, deep.

BEAK: Of medium size, stout, at base, well curved.

EYES: Full.

COMB: Rose; large, square in front, free from hollow center, uniform on each side; top covered with small points, firm and even on head without inclining to either side, terminating at rear in a well developed and straight spike.

WATTLES AND EAR-LOBES: Of medium size.

NECK: Rather long, with full hackle flowing well over shoulders.

WINGS: Large, folded well against body.

BACK: Of medium length, sloping straight to tail; saddle feathers, long and sweeping.

TAIL: Full, well expanded, carried at an angle of fifty degrees above, the horizontal (see illustrations, figures 39 and 41); sickles, long, well curved; coverts, abundant.

BREAST: Broad, deep, prominent.

BODY AND FLUFF: Body, long, rounded, broadest in front and tapering to rear; fluff, rather short.

LEGS AND TOES: Thighs, of medium length, well developed; shanks, rather long; toes, straight, well spread.

SHAPE OF FEMALE.

HEAD: Short, deep.

BEAK: Of medium size, stout at base, well curved.

EYES: Full.

COMB: Similar to that of male, but smaller.

WATTLES AND EAR-LOBES: Wattles, of medium size, well rounded. Ear-lobes, of medium size.

NECK: Rather long, full feathered.

WINGS: Large, well folded against body.

BACK: Long, straight, sloping slightly to tail.

TAIL: Long, full, well expanded, carried at an angle of forty-five degrees above the horizontal. (See illustration, figure 39.)

BREAST: Broad, prominent.

BODY AND FLUFF: Body, long, rounded, deep; fluff, rather short.

LEGS AND TOES: Thighs, of medium length, well developed; shanks, of medium length; toes, straight, well spread.

COLOR OF MALE.

HEAD: Plumage, rich, dark red.

BEAK: Horn.

EYES: Reddish-bay.

COMB, FACE, WATTLES AND EAR-LOBES: Bright red.

NECK: Hackle, blue-black, each feather edged with red, the entire hackle shading off to black at base; plumage in front of hackle, black.

WINGS: Bows, mahogany red; coverts, rich, deep brown, each feather ending with a black spangle shaped like a half-moon, forming double black bars across wings; primaries, dull black; secondaries, upper web black, lower web black with a broad edging of brown, each feather ending with a bluish-black spangle, shaped like a half-moon.

BACK: Rich red and black; saddle feathers, rich, dark red, with a bluish-black stripe extending down middle of each feather.

TAIL: Black; sickles and coverts, greenish-black.

BREAST: Lustrous black.

BODY AND FLUFF: Black.

LEGS AND TOES: Thighs, black; shanks and toes, slate or leaden-blue.

UNDER-COLOR OF ALL SECTIONS: Dark leaden-blue.

COLOR OF FEMALE.

HEAD: Plumage, brown.

BEAK: Horn.

EYES: Reddish-bay.

COMB, FACE, WATTLES AND EAR-LOBES: Bright red.

NECK: Black, each feather laced with golden-bay; feathers in front of neck, same as breast.

WINGS: Bows, rich brown, each feather ending with a bluish-black spangle, shaped like a half-moon; coverts, similar to bows; primaries, dull black, with a narrow edging of brown on lower webs; secondaries, black, lower webs with a broad edging of brown, each feather ending with a bluish-black spangle, shaped like a half-moon.

BACK: Rich brown, each feather ending with a bluish-black spangle, shaped like a half-moon.

TAIL: Black; coverts, brown, each feather ending with a bluish-black spangle, shaped like a half-moon.

BREAST: Rich brown, each feather ending with a bluish-black spangle, shaped like a half-moon.

BODY AND FLUFF: Body, similar to that of breast, but shading off lighter on under parts; fluff, black, powdered with brown.

LEGS AND TOES: Thighs, light brown; shanks and toes, slate or leaden-blue.

UNDER-COLOR OF ALL SECTIONS: Dark, leaden-blue.

ORPINGTONS.

The Orpingtons are of English origin, and since their introduction in their native land, have been one of the most popular breeds of fowls. They are stately in appearance; with rather long, round, deep bodies, full breasts, and broad backs; abundance of hackle and saddle feathers on the male giving the appearance of a rather short back; legs, rather short and well apart, and shanks large and nearly round. The breeders of these varieties should strive to maintain the true type and color of plumage; to hold the Buffs to one even shade of rich golden-buff, free from white or black; the Blacks to a lustrous greenish-black throughout; the Whites, to a pure white in all sections; the Blues, to a clear, even, medium shade of slaty-blue, each feather having a clear and well defined lacing of a darker blue.

Orpington Disqualifications.

Positive white in ear-lobes, covering more than one-third the surface; yellow beak, shanks or skin. (See general disqualifications.)

STANDARD WEIGHTS.

Cock ....10 lbs.   Hen ....8 lbs.
Cockerel .... 8½ lbs.   Pullet ....7 lbs.

SHAPE OF MALE.

HEAD: Rather long, broad, deep.
BEAK: Short, stout, regularly curved.
EYES: Large, oval.
COMB: Single; rather large, set firmly on head, perfectly straight and upright; with five well defined points, those at front and rear smaller than those in the middle; fine in texture; blade closely following shape of head.
WATTLES AND EAR-LOBES: Wattles, of medium size, well rounded at lower edges. Ear-lobes, medium size, oblong, smooth.
NECK: Rather short, well arched, with abundant hackle.
WINGS: Of medium size, well folded; fronts, well covered by breast feathers; points well covered by saddle feathers.
BACK: Broad, flat at shoulders, of medium length, width car-

ried well back to base of tail; rising with a full concave sweep to tail; saddle feathers, of medium length, abundant.

TAIL: Moderately long, well spread, carried at an angle of forty-five degrees from the horizontal (see illustration, figure 39), forming no apparent angle with back where those sections join; sickles of medium length, spreading laterally beyond main tail feathers; smaller sickles and tail-coverts, of medium length, nicely curved, sufficiently abundant to cover main tail feathers.

BREAST: Broad, deep, well rounded.

BODY AND FLUFF: Body, broad, deep, rather long, straight, extending well forward; fluff, moderately full.

LEGS AND TOES: Thighs, large, rather short, covered with soft feathers; shanks, rather short, set well apart, stout in bone, smooth; toes, of medium length, straight, strong, well spread.

SHAPE OF FEMALE.

HEAD: Rather large, broad, deep.

BEAK: Short, stout, regularly curved.

EYES: Large, oval.

COMB: Single, of medium size; set firmly on head, perfectly straight and upright; with five well defined points, those in front and rear smaller than the middle ones; fine in texture.

WATTLES AND EAR-LOBES: Wattles, of medium length, fine in texture. Ear-lobes, of medium size, oblong.

NECK: Rather short, well arched, nicely tapering to head, having moderately full plumage.

WINGS: Of medium size, well folded.

BACK: Broad, moderately long, width carried well back to base of tail; rising with concave sweep to tail.

TAIL: Moderately long, well spread, carried at an angle of forty-five degrees from the horizontal (see illustration, figure 39); tail-coverts, abundant.

BREAST: Broad, deep, well rounded.

BODY AND FLUFF: Body, long, broad, deep, straight, extending well forward; fluff, moderately full.

LEGS AND TOES: Thighs, large, rather short, covered with soft feathers; shanks rather short, stout, set well apart; toes, of medium length, straight, strong, and well spread.

BUFF ORPINGTON MALE

190

BUFF ORPINGTON FEMALE

BUFF ORPINGTON MALE

BUFF ORPINGTON FEMALE

BUFF ORPINGTONS.

Disqualifications.

Shanks other than white or pinkish-white. (See general and Orpington disqualifications.)

COLOR OF MALE.

BEAK: White, or pinkish-white.

EYES: Reddish-bay.

COMB, FACE, WATTLES AND EAR-LOBES: Bright red.

SHANKS AND TOES: White, or pinkish-white.

PLUMAGE: Surface throughout an even shade of rich, golden-buff, free from shafting or mealy appearance; the head, neck, hackle, back, wing-bows and saddle richly glossed. Under-color, a lighter shade of buff, free from foreign color. Different shades of buff in two or more sections is a serious defect. A harmonious blending of buff in all sections is most desirable.

COLOR OF FEMALE.

BEAK: White, or pinkish white.

EYES: Reddish-bay.

COMB, FACE, WATTLES AND EAR-LOBES: Bright red.

SHANKS AND TOES: White or pinkish-white.,

PLUMAGE: Surface throughout an even shade of rich, golden buff, free from shafting or mealy appearance, the head and neck plumage showing a luster of the same shade as the rest of the plumage. Under color a lighter shade, free from foreign color. Different shades of buff in two or more sections is a serious defect. A harmonious blending of buff is most desirable.

BLACK ORPINGTONS.

Disqualifications.

One-half inch or more of white in any part of plumage. (See general and Orpington disqualifications.)

COLOR OF MALE AND FEMALE.

BEAK: Black.

EYES: Black, or dark brown.

COMB, FACE, WATTLES AND EAR-LOBES: Bright red.

SHANKS AND TOES: Black; web and bottom of toes, pinkish-white.

PLUMAGE: Surface, lustrous greenish-black throughout.

UNDER-COLOR OF ALL SECTIONS: Dull black.

WHITE ORPINGTONS.

Disqualifications.

Feathers other than white in any part of plumage; shanks other than white or pinkish-white. (See general and Orpington disqualifications.)

COLOR OF MALE AND FEMALE.

BEAK: Pinkish-white.

EYES: Reddish-bay.

COMB, FACE, WATTLES AND EAR-LOBES: Bright red.

SHANKS AND TOES: White or pinkish-white.

PLUMAGE: Web, fluff and quills of feathers in all sections, pure white.

BLUE ORPINGTONS.

Disqualifications.

Red or positive white in color of plumage. (See general and Orpington disqualifications.)

COLOR OF MALE.

HEAD: Plumage, dark slaty-blue.

BEAK: Horn.

EYES: Dark brown.

COMB, FACE AND WATTLES: Bright red.

EAR-LOBES: Red.

NECK: Very dark, lustrous blue.

WINGS: Bows, very dark, lustrous blue; coverts, a clear, even medium shade of slaty-blue, having a well defined lacing of darker blue; primaries, a clear, even, medium shade of slaty-blue; secondaries, inner webs a clear, even shade of slaty-blue, outer webs, slaty-blue, each feather having a clear, well defined lacing of darker blue.

BACK: Very dark, lustrous blue.

TAIL: Sickles, very dark, lustrous blue; tail coverts, lustrous blue; main tail feathers, a clear, even slaty-blue, each feather having a well defined lacing of darker blue.

BREAST: A clear, even, medium shade of slaty-blue, each feather having a clear and well defined lacing of darker blue.

BODY AND FLUFF: Body, a clear, even, medium shade of slaty-blue, each feather having a clear, well defined lacing of darker blue; fluff, slaty-blue.

WHITE ORPINGTON MALE

194

WHITE ORPINGTON FEMALE

195

LEGS AND TOES: Thighs, a clear, even shade of slaty-blue, each feather having a well defined lacing of darker blue; shanks and toes, leaden-blue.

UNDER-COLOR OF ALL SECTIONS: Slaty-blue.

COLOR OF FEMALE.

HEAD: Plumage, slaty-blue.

BEAK: Horn.

EYES: Dark brown.

COMB, FACE AND WATTLES: Bright red.

EAR-LOBES: Red.

NECK: Slaty-blue, laced with a darker blue.

WINGS: Primaries, a clear, even, medium shade of slaty-blue; remainder of wing an even shade of slaty-blue, darker than that of primaries; feathers in all sections, except primaries, having clear and well defined lacings of darker blue.

BACK: Slaty-blue, each feather having a clear and well defined lacing of darker blue.

TAIL: Slaty-blue, each feather laced with a darker blue.

BREAST: Slaty-blue, each feather having a clear and well defined lacing of darker blue.

BODY AND FLUFF: Body, slaty-blue, each feather having a clear and well defined lacing of darker blue; fluff, slaty-blue.

LEGS AND TOES: Thighs, slaty-blue, each feather having a clear and well defined lacing of darker blue; shanks and toes, leaden-blue.

UNDER-COLOR OF ALL SECTIONS: Slaty-blue.

CORNISH. •

The Cornish fowl is a composite of several different blood-lines of which the strongest and most prominent is the Aseel. From this breed comes the color of the Dark variety and the great luster of plumage; also, the pea comb, short and close plumage, thick legs, large thighs, deep and broad breast, great width of back at shoulders, comparatively short neck and projecting brows. The fowls are sturdy in appearance; the thighs are stout and muscular; and the breast is very broad and rounded at the sides. The carriage is upright, the shoulders being carried high and the stern low. The body between the thighs should be very wide. The Dark variety was perfected in England, the White and the White-Laced Red in America.

Dark and White Cornish.

STANDARD WEIGHTS.

Cock ....10 lbs.   Hen ....7½ lbs.
Cockerel .... 8 lbs.   Pullet 6   lbs.

White-Laced Red Cornish.

STANDARD WEIGHTS.

Cock ....8 lbs.   Hen ....6 lbs.
Cockerel ....7 lbs.   Pullet ....5 lbs.

SHAPE OF MALE.

HEAD: Short, deep and broad, indicating great vigor and strong constitution; the crown projecting over the eyes.

BEAK: Short, stout, well curved.

EYES: Full, with bold and fearless expression, not sunken in the sense of being close together.

COMB AND FACE: Comb, pea; small, firmly and closely set upon head. Face, rather coarse in texture.

WATTLES AND EAR-LOBES: Wattles, small, even, smooth in texture. Ear-lobes, small, smooth in texture.

NECK: Medium in length, slightly arched; throat, full, dotted with small feathers.

WINGS: Short and muscular, closely folded; fronts standing out prominently from body and shoulders; points, slightly rounded at extreme ends when folded, closely tucked at ends and held on a line with lower tail-coverts.

BACK: Medium in length, top line of back slightly convex, sloping downward from base of neck to tail, and sloping slightly from each side of backbone, well filled in at base of neck; hip bones very wide apart; very broad across shoulders, carrying its width well back to a line with thighs, showing good width between wings and then narrowing to tail.

TAIL: Short and closely folded, carried at or slightly below the horizontal. (See illustration, figure 39.)

BREAST: Broad and deep, well rounded at sides, projecting beyond wing-fronts when specimen is standing erect.

BODY AND FLUFF: Body, well rounded at sides, long and straight; fluff, well tucked up.

LEGS AND TOES: Thighs, of medium length, round, muscular, set well apart; shanks, short and stout in bone; toes, long, strong and straight, well spread.

PLUMAGE: Short, narrow, close.

STATION AND CARRIAGE: Station, low; carriage, erect, upright, indicating great vigor.

SHAPE OF FEMALE.

HEAD: Short, deep, broad, indicating great vigor and strong constitution; crown projecting over the eyes.

BEAK: Well curved.

EYES: Full, with bold and fearless expression, not sunken in the sense of being close together.

COMB AND FACE: Comb, pea; small and closely set on head. Face, rather coarse in texture.

WATTLES AND EAR-LOBES: Wattles, small, even, smooth in texture. Ear-lobes, small, smooth in texture.

NECK: Medium in length, slightly arched; throat, full, dotted with small feathers.

WINGS: Short and muscular, closely folded; fronts standing out prominently from body and shoulders; points, slightly rounded at extreme ends when folded, closely tucked at ends and held on a line with lower tail-coverts.

BACK: Medium in length; top line of back slightly convex, sloping downward from base of neck to tail, and sloping slightly from each side of backbone; well filled in at base of neck; hip bones very wide apart; very broad across shoulders, carrying its width well back to a line with the thighs, showing good width between wings and then narrowing to tail.

TAIL: Short and closely folded, carried at or slightly below the horizontal. (See illustration, figure 39.)

BREAST: Broad and deep, well rounded at sides, projecting beyond wing-fronts when specimen is standing erect.

BODY AND FLUFF: Body, well rounded at sides, long and straight; fluff, well tucked up.

LEGS AND TOES: Thighs, of medium length, round, muscular, set well apart; shanks, short and stout; toes, long, strong, straight, well spread.

PLUMAGE: Short, narrow and close.

STATION AND CARRIAGE: Station, low; carriage, erect, upright, indicating great vigor.

DARK CORNISH.

Disqualifications.

Solid white, blue, or black shanks. (See general disqualifications.)

COLOR OF MALE.

HEAD: Greenish-black.

BEAK: Yellow.

EYES: Yellow, approaching pearl.

COMB, FACE, WATTLES AND EAR-LOBES: Bright red.

NECK: Hackle, lustrous, greenish-black; shafts, red; plumage, other than hackle, lustrous greenish-black; shafts, red.

WINGS: Fronts, greenish-black; bows, lustrous, greenish-black and dark red intermixed, the black greatly predominating; coverts, forming wing-bars, lustrous, greenish-black; primaries, black except a narrow edging of bay on outer webs; secondaries, upper webs black, lower webs, one-third black next to shaft, the remainder bay.

BACK: Lustrous greenish-black and dark red intermixed, the black greatly predominating; saddle feathers, like back in color, but with a somewhat larger proportion of dark red.

TAIL: Black; sickles and coverts, lustrous greenish-black.

BREAST: Lustrous greenish-black.

BODY AND FLUFF: Black.

LEGS AND TOES: Thighs, black; shanks and toes, yellow.

UNDER-COLOR OF ALL SECTIONS: Dark slate.

DARK CORNISH MALE

DARK CORNISH FEMALE

COLOR OF FEMALE.

HEAD: Plumage, greenish-black.

BEAK: Yellow.

EYES: Yellow, approaching pearl.

COMB, FACE, WATTLES AND EAR-LOBES: Bright red.

NECK: Upper portion, lustrous-black, with bay shaft to each feather; feathers in front of neck, bay, approaching mahogany, each feather having two pencilings of lustrous black, the penciling following contour of feathers.

WINGS: Bows and coverts, bay approaching mahogany, each feather having two pencilings of lustrous black, the pencilings following contour of feather; primaries, black, except a narrow edging of irregularly penciled bay on outer part of webs; secondaries, upper webs black, lower webs black next to shaft of feather, with a broad margin of irregularly penciled bay.

BACK: Bay, approaching mahogany, each feather having two pencilings of black, the pencilings following the contour of feather.

TAIL: Main feathers, black, except the two upper feathers, which are irregularly penciled with bay; coverts, bay, approaching mahogany, each feather having two pencilings of lustrous black, the pencilings following the contour of feather.

BREAST: Bay, approaching mahogany, each feather having two pencilings of lustrous black, the pencilings following contour of feather.

BODY AND FLUFF: Body, bay, approaching mahogany, each feather having two pencilings of lustrous black, the pencilings following contour of feather; fluff, black, or black tinged with bay.

LEGS AND TOES: Thighs, black, more or less penciled with bay; shanks and toes, yellow.

UNDER-COLOR OF ALL SECTIONS: Dark slate.

WHITE CORNISH.

Disqualifications.

Red, buff, or positive black in any part of plumage; solid green or white shanks. (See general disqualifications.)

COLOR OF MALE AND FEMALE.

BEAK: Yellow.

EYES: Yellow, approaching pearl.

COMB, FACE, WATTLES AND EAR-LOBES: Bright red.

SHANKS AND TOES: Yellow.

PLUMAGE: Web, fluff and quill of feathers in all sections, pure white.

WHITE-LACED RED CORNISH.

Disqualifications.

Solid white, blue or black shanks. (See general disqualifications.)

COLOR OF MALE.

HEAD: Plumage, rich, bright red, each feather tipped with white.

BEAK: Yellow.

EYES: Yellow, approaching pearl.

COMB, FACE, WATTLES AND EAR-LOBES: Bright red.

NECK: Bright, rich red, each feather laced with silvery white; plumage in front of neck, same as breast.

WINGS: Fronts, shoulders and bows, bright, rich red, each feather regularly laced with a narrow lacing of silvery white; coverts, bright, rich red, forming wing-bars, regularly laced with white; primaries, bright, rich red, with well defined, regular lacings of white; secondaries, bright, rich red; with well defined regular lacings of white; flight coverts, red, laced with white.

BACK: Bright, rich red, each feather ending with silvery white approaching the letter V in shape; saddle feathers, prominent, each feather laced with silvery white, the texture of the feather giving a rayed appearance.

TAIL: White, with shaft and extreme center red; sickles and coverts, white, with red shaft and center.

BREAST: Bright, rich red, each feather laced with a narrow, regular lacing of white, following shape of web to fluff.

BODY AND FLUFF: Bright, rich red, each feather regularly laced with a narrow lacing of white throughout.

LEGS AND TOES: Thighs, bright, rich red, each feather regularly laced with white; shanks and toes, bright yellow.

UNDER-COLOR OF ALL SECTIONS: White.

COLOR OF FEMALE.

HEAD: Plumage, bright, rich red, each feather laced with white.

BEAK: Yellow.

EYES: Yellow, approaching pearl.

COMB, FACE, WATTLES AND EAR-LOBES: Bright red.

NECK: Bright, rich red, each feather regularly laced with white; feathers in front of neck, same as breast.

WINGS: Bright, rich red, with shoulders, fronts, bows and

WHITE-LACED RED CORNISH MALE

WHITE-LACED RED CORNISH FEMALE

wing-bars regularly laced with a narrow lacing of white; primaries, bright, rich red, ending with white and well up on lower edge; secondaries, bright, rich red, with well defined, regular lacings of white.

BACK: Bright, rich red throughout, with each feather, from cape to tail, regularly laced with a narrow white lacing conforming perfectly to shape of feather; free from mossiness or barrings.

TAIL: Red, each feather laced with white, the white being wider at extremity of feather.

BREAST: Bright, rich red, with each feather regularly laced with a narrow lacing of white, conforming perfectly to shape of feather; free from mossiness or barrings.

BODY AND FLUFF: Bright, rich red, each feather regularly laced with white.

LEGS AND TOES: Thighs, red, each feather regularly laced with white; shanks and toes, bright yellow.

UNDER-COLOR OF ALL SECTIONS: White.

SUSSEX.

The Sussex fowl had its origin in England and takes its name from the County of Sussex. The body should be long, broad at shoulders, and deep, presenting an apparently oblong profile when viewed from the side. White legs, flesh and skin are characteristics of the breed. Reasonable and not excessive weights are required and therefore standard weight should be attained.

STANDARD WEIGHTS.

Cock 9   lbs.   Hen ....7 lbs.
Cockerel ....7½ lbs.   Pullet ....6 lbs.

Disqualifications.

White in ear-lobes covering more than one-third of surface; yellow skin, shanks or feet. (See general disqualifications.)

SHAPE OF MALE.

HEAD: Moderately large, broad, medium in length.

BEAK: Stout, rather short, slightly curved.

EYES: Full, oval.

COMB: Single; medium in size, set closely to head, perfectly straight and upright, having five well defined points, those at front and rear being smaller than those in the middle; fine in texture, blade free from serrations, and following curve of neck.

WATTLES AND EAR-LOBES: Wattles of medium size, equal in length, well rounded at lower edges. Ear-lobes, oval, medium in size.

NECK: Of medium length, slightly arched, having abundant hackle, flowing well over shoulders.

WINGS: Rather long, well folded, carried close to body; points well covered with saddle feathers.

BACK: Long, flat, and broad its entire length, sloping slightly to tail; saddle feathers, of medium length, abundant.

TAIL: Of medium length, well spread, main tail feathers carried at an angle of forty-five degrees above the horizontal (see illustration, fig. 39), thus increasing the apparent length of the fowl; sickles of medium length, well curved, extending slightly beyond main tail feathers; smaller sickles and **tail-coverts** of medium length.

BREAST: Deep, full.

BODY AND FLUFF: Body, long, broad, deep; fluff, moderately full, fitting closely to body.

LEGS AND TOES: Thighs, stout, of medium length; shanks, of medium length, straight, stout, set well apart; toes, straight, of medium length, well spread.

SHAPE OF FEMALE.

HEAD: Moderately large, broad, medium in length.

BEAK: Stout, rather short, slightly curved.

EYES: Full, oval.

COMB: Single; rather small in size, set closely to head, perfectly straight and upright, having five well defined points, those at front and rear being smaller than those in the middle; fine in texture.

WATTLES AND EAR-LOBES: Wattles, small, equal in length, well rounded at lower edges. Ear-lobes, medium in size, oval.

NECK: Of medium length, nicely curved and tapering; neck feathers, moderately full, flowing well over shoulders with no apparent break at juncture of neck and back.

WINGS: Medium in length, well folded, carried close to body.

BACK: Long, flat, and broad its entire length, sloping slightly to tail.

TAIL: Of medium length, fairly well spread, carried at an angle of forty degrees above the horizontal (see illustration, figure 39; tail-coverts, well developed.

BREAST: Broad, full, deep.

BODY AND FLUFF: Body, long, broad, deep; fluff, moderately full, fitting closely to body.

LEGS AND TOES: Thighs, stout, of medium length; shanks, of medium length, straight, set well apart; toes, straight, of medium length, well spread.

SPECKLED SUSSEX.

COLOR OF MALE.

HEAD: Plumage, lustrous reddish-brown, each feather tipped with white, a narrow black bar dividing the white from balance of feather.

BEAK: Horn.

EYES: Reddish-bay.

COMB, FACE, WATTLES AND EAR-LOBES: Red.

NECK: Lustrous reddish-brown, with black stripe extending lengthwise to lower part of each feather, which should terminate with a white tip at its lower extremity; plumage in front of neck, same as breast.

WINGS: Bows, lustrous reddish-brown, each feather tipped with white, a narrow black bar dividing the white from balance of feather; coverts, reddish-brown, each feather tipped with a large white spangle, a black bar dividing the white from balance of feather; primaries, white and black; secondaries, outer web, bay, edged with white, inner web, black, edged with white, each feather ending with a white spangle.

BACK: Lustrous reddish-brown, with black stripe extending lengthwise to lower part of feather, which should terminate with a white tip at its lower extremity.

TAIL: Main tail feathers, black and white; sickles, coverts and smaller coverts, glossy, greenish-black, tipped with white.

BREAST: Reddish-brown, each feather tipped with white, a black bar dividing the white from balance of feather.

BODY AND FLUFF: Reddish-brown, each feather tipped with white, a black bar dividing the white from balance of feather.

LEGS AND TOES: Thighs, reddish-brown, each feather tipped with white, a narrow black bar dividing the white from balance of feather; shanks and toes, white.

UNDER-COLOR OF ALL SECTIONS: Slate, shading into white at base.

COLOR OF FEMALE.

HEAD: Plumage, reddish-brown, each feather tipped with a white spangle, a narrow black bar dividing the white from balance of feather.

BEAK: Horn.

EYES: Reddish-bay.

COMB, FACE, WATTLES AND EAR-LOBES: Red.

SPECKLED SUSSEX MALE

SPECKLED SUSSEX FEMALE

NECK: Reddish-brown, each feather tipped with a white spangle, a narrow black bar dividing the white from balance of feather; feathers in front of neck, same as breast.

WINGS: Bows and coverts, reddish-brown, each feather tipped with a small white spangle, a narrow black bar dividing the white from balance of feathers; primaries, black and white; secondaries, outer web, bay, edged with white, inner web, black, edged with white, each feather ending with a white spangle.

BACK: Reddish-brown, each feather tipped with a small white spangle, a narrow black bar dividing white from balance of feather.

TAIL: Main tail, black mottled with brown, each feather tipped with white; coverts, reddish-brown, each feather marked with a crescentic black bar, near lower extremity, and tipped with white.

BREAST: Reddish-brown, each feather tipped with white, a black bar dividing the white from balance of feather.

BODY AND FLUFF: Reddish-brown, each feather tipped with white, a black bar dividing the white from balance of feather.

LEGS AND TOES: Thighs, reddish-brown, each feather tipped with a white spangle, a narrow black bar dividing the white from balance of feather; shanks and toes, white.

UNDER-COLOR OF ALL SECTIONS: Slate, shading into white at base of feathers.

RED SUSSEX.

Disqualifications.

One or more white feathers showing in outer plumage. (See general and Sussex disqualifications.)

COLOR OF MALE.

HEAD: Plumage, mahogany red.

BEAK: Horn.

EYES: Reddish-bay.

COMB, FACE, WATTLES AND EAR-LOBES: Red.

NECK: Lustrous mahogany red, plumage in front of hackle, mahogany red.

WINGS: Bows, lustrous mahogany red; coverts, red; primaries, upper web red, lower web black with narrow edging of red, only sufficient to prevent the black from showing on surface when wings are folded in natural position; primary coverts, black; secondaries, lower web red, sufficient to give a red wing-bay; the remainder of each feather black.

BACK: Lustrous mahogany red.

TAIL: Main tail and sickle feathers, black or greenish-black; tail coverts, mainly black, becoming red as saddle is approached.

BREAST: Rich mahogany red.

BODY AND FLUFF: Rich mahogany red.

SHANKS AND TOES: White.

UNDER-COLOR OF ALL SECTIONS: Slate, shading into red at base.

COLOR OF FEMALE.

HEAD: Plumage, mahogany red.

BEAK: Horn.

EYES: Reddish-bay.

COMB, FACE, WATTLES AND EAR-LOBES: Red.

NECK: Rich mahogany red.

WINGS: Bows, rich mahogany red; coverts, red; primaries, upper webs red, lower webs black with narrow edging of red, only sufficient to prevent black from showing on surface when wings are folded; primary coverts, black; secondaries, lower web, red, the remainder of each feather black, forming a red bay.

BACK: Rich mahogany red.

TAIL: Black, except the two top feathers which may be edged with red.

BREAST: Rich mahogany red.

BODY AND FLUFF: Rich mahogany red.

SHANKS AND TOES: White.

UNDER-COLOR OF ALL SECTIONS: Slate, shading into red at base.

POLISH.

Breed   *Varieties*

White-Crested Black
Bearded Golden
Bearded Silver
POLISH ....⎬ Bearded White
Buff-Laced
Non-Bearded Golden
Non-Bearded Silver
Non-Bearded White

SCALE OF POINTS FOR THE POLISH CLASS.

Symmetry 4
Size 4
Condition 4
Comb 2
Crest—Shape 10, Color 5...................................... 15
Head—Shape 2, Color 2.. 4
Beak—Shape 2, Color 2.. 4
Eyes—Shape 2, Color 2.. 4
Wattles and Ear-Lobes—Shape 2, Color 2, Beard* 4......... 8
Neck—Shape 3, Color 3.. 6
Wings—Shape 4, Color 6....................................... 10
Back—Shape 4, Color 4.. 8
Tail—Shape 4, Color 5 9
Breast—Shape 4, Color 4...................................... 8
Body and Fluff—Shape 3, Color 3............................. 6
Legs and Toes—Shape 2, Color 2.............................. 4
                                                        ——
                                                        100

*Omit "Beard" in the non-bearded varieties and give the four points to crest, two for shape and two for color.

POLISH.

Polish, as pure bred fowls, are so long established that they antedate the history of standard bred fowls. We find them mentioned as early as the 16th century and as a pure bred race, they are, no doubt, much older than even that date would indicate. Among all the domestic races of fowls, Polish take high rank as ornamental and beautiful. For a breed of moderate size, Polish are quiet and docile. Most conspicuous of all Polish characteristics is the crest. The range of color in the several varieties and the peculiar markings of the laced variety furnish the breeders rare opportunities of testing their skill.

SHAPE OF MALE.

HEAD: Large with a pronounced protuberance on top of skull.

BEAK: Of medium length, slightly curved.

NOSTRILS: Large, with crown elevated above the ordinary curved line of beak.

EYES: Large, full.

COMB AND CREST: Comb, V-shaped, of small size, the smaller, the better; set evenly on head, retreating into crest; natural absence of comb is preferred. Crest, very large, profuse, rising well in front so as not to obstruct the sight, and falling over on both sides and in rear in a regular, even mass, composed of feathers similar in shape and texture to those of hackle.

BEARD: (In bearded varieties) thick and full, running back of eyes in a graceful curve.

WATTLES AND EAR-LOBES: Wattles, of equal length, thin, well rounded on lower edges. Ear-lobes, small, smooth.

NECK: Of medium length, slightly arched, with abundant hackle flowing well over shoulders.

WINGS: Large, well folded.

BACK: Straight, wide across the shoulders tapering to tail; saddle feathers, abundant, with slight concave sweep near base of tail.

TAIL: Large, well expanded, carried at an angle of forty degrees above the horizontal (see illustration, figure 39); sickles, and coverts, abundant and covering main tail feathers.

BREAST: Full, prominent.

BODY AND FLUFF: Body, of medium length, moderately full, tapering from front to rear; fluff, rather short.

LEGS AND TOES: Thighs, of medium length; shanks, of medium length, slender; toes, straight.

SHAPE OF FEMALE.

HEAD, BEAK, NOSTRILS, EYES, COMB, WATTLES AND EAR-LOBES: Similar to those of male, but smaller.

CREST: Large, profuse, globular, rising well in front, regular and unbroken.

BEARD: (In bearded varieties) same as male.

NECK: Medium length, full at base, tapering to head.

WINGS: Large, well folded.

BACK: Straight, broadest at shoulders, tapering, with concave sweep near base of tail.

TAIL: Large, broad, well expanded, carried at an angle of forty degrees above the horizontal. (See illustration, figure 39.)

BREAST: Full, round, prominent.

BODY AND FLUFF: Body, medium in length, moderately full; fluff, short.

LEGS AND TOES: Thighs, of medium length; shanks, of medium length, slender; toes, straight.

WHITE-CRESTED BLACK POLISH.

Disqualifications.

Positive white in any part of plumage extending over half an inch, except in crest; shanks other than blue or dark leaden-blue. (See general disqualifications.)

COLOR OF MALE AND FEMALE.

HEAD: Face, red.

BEAK: Bluish-black.

EYES: Reddish-bay.

COMB AND CREST: Comb, red. Crest, white, a narrow band of black feathers at base of crest in front is allowable, but the fewer the better.

WATTLES AND EAR-LOBES: Wattles, red; ear-lobes, white.

SHANKS AND TOES: Bluish-black.

PLUMAGE, EXCEPT CREST: Lustrous greenish-black.

UNDER-COLOR OF ALL SECTIONS EXCEPT CREST: Dull black.

BEARDED GOLDEN POLISH.

Disqualifications.

Shanks other than blue or slaty-blue. (See general disqualifications.)

COLOR OF MALE.

HEAD: Face, red.

BEAK: Dark horn.

EYES: Reddish-bay.

COMB AND CREST: Comb, red. Crest, golden bay, laced with black.

BEARD: Golden-bay, laced with black.

WATTLES AND EAR-LOBES: Wattles, red; ear-lobes, white.

NECK: Golden-bay, each feather laced with black; plumage in front of neck, same as breast.

WINGS: Bows, golden-bay, each feather laced with black; coverts, golden-bay, each feather laced with black, lacing widest at end, forming two well defined wing bars; primaries, golden-bay, each feather ending with black, the black tapering to a point on lower edge; secondaries, golden-bay, with well defined black lacings.

BACK: Golden-bay, each feather laced with black; saddle feathers, abundant, each feather laced with black, the texture of feathers giving a rayed appearance.

TAIL: Golden-bay, each feather laced with black; lacing widest at end; sickles and coverts, golden-bay each feather laced with black, lacing widest at end.

BREAST: Golden-bay, free from mossiness, each feather laced with black, the lacing being proportionate to size of feather.

BODY AND FLUFF: Body, golden-bay, each feather laced with black; fluff, bay, tinged with black.

LEGS AND TOES: Thighs, bay, each feather laced with black; shanks and toes, slaty-blue.

UNDER-COLOR OF ALL SECTIONS: Slate.

COLOR OF FEMALE.

HEAD: Face, red.

BEAK: Dark horn.

EYES: Reddish-bay.

COMB AND CREST: Comb, red. Crest, in pullets, black laced with bay, which after first moult should be golden-bay laced with black.

217

WHITE-CRESTED BLACK POLISH MALE

WHITE-CRESTED BLACK POLISH FEMALE

BEARD: Golden-bay, heavily laced with black.

WATTLES AND EAR-LOBES: Wattles, red. Ear-lobes, white.

NECK: Golden-bay, each feather laced with black; feathers in front of neck same as breast.

WINGS: Bows, golden-bay, each feather laced with black; coverts, golden-bay, each feather laced with black, the black growing wider at the extremity, forming two distinctly laced bars across wings; primaries, golden-bay, each feather ending with black, the black tapering to a point on lower edge; secondaries, golden-bay with well defined black lacings.

BACK: Golden-bay, each feather laced with black.

TAIL: Golden-bay, each feather laced with black, the black being wider at outer end of feather.

BREAST: Golden-bay, each feather laced with black.

BODY AND FLUFF: Body, golden-bay, each feather laced with black; fluff, a lighter shade of bay, tinged with black.

LEGS AND TOES: Thighs, bay, each feather laced with black; shanks and toes, slaty-blue.

UNDER-COLOR OF ALL SECTIONS: Slate.

BEARDED SILVER POLISH.

Disqualifications.

Shanks other than blue or slaty-blue. (See general disqualifications.)

COLOR OF MALE.

HEAD: Face, red.

BEAK: Dark horn.

EYES: Reddish-bay.

COMB AND CREST: Comb, red. Crest, white, laced with black.

BEARD: White, laced with black.

WATTLES AND EAR-LOBES: Wattles, red. Ear-lobes, white.

NECK: White, each feather laced with black; plumage in front of neck, same as breast.

WINGS: Bows, white, each feather laced with black; coverts, white, each feather laced with black, lacing widest at end, forming two well defined wing bars; primaries, white, each feather ending with black, the black tapering to a point on lower edges; secondaries, white, with well defined black lacings.

BACK: White, each feather laced with black; saddle feathers, white, each feather laced with black, the texture of feathers giving a rayed appearance.

TAIL: White, each feather laced with black, lacing widest at ends; sickles and coverts, white, each feather laced with black, lacings widest at ends.

BREAST: White, free from mossiness, each feather laced with black, the lacing being proportionate to size of feather.

BODY AND FLUFF: Body, white, each feather laced with black; fluff, white, tinged with black.

LEGS AND TOES: Thighs, white laced with black; shanks and toes, slaty-blue.

UNDER-COLOR OF ALL SECTIONS: Slate.

COLOR OF FEMALE.

HEAD: Face, red.

BEAK: Dark horn.

EYES: Reddish-bay.

COMB AND CREST: Comb, red. Crest, in pullets, black laced with white, which after first moult should be white laced with black.

BEARD: White, heavily laced with black.

WATTLES AND EAR-LOBES: Wattles, red. Ear-lobes, white.

NECK: White, each feather laced with black; feathers in front of neck, same as breast.

WINGS: Bows, white, each feather laced with black; coverts, white, each feather laced with black, the black growing wider at the extremity, forming two distinctly laced bars across wing; primaries, white, each feather ending with black, the black tapering to a point on lower edge; secondaries, white, with well defined black lacings.

BACK: White, each feather laced with black.

TAIL: White, each feather laced with black, the black being wider at outer end of feather.

BREAST: White, each feather laced with black.

BODY AND FLUFF: Body, white, each feather laced with black; fluff, white tinged with black.

LEGS AND TOES: Thighs, white, each feather laced with black; shanks and toes, slaty-blue.

UNDER-COLOR OF ALL SECTIONS: Slate.

BEARDED SILVER POLISH MALE

BEARDED SILVER POLISH FEMALE

223

BEARDED WHITE POLISH.

Disqualifications.

Shanks other than blue or slaty-blue. (See general disqualifications.)

COLOR OF MALE AND FEMALE.

BEAK: Dark horn.
EYES: Reddish-bay.
COMB, FACE AND WATTLES: Red.
EAR-LOBES: White.
SHANKS AND TOES: Slaty-blue.
PLUMAGE: Web, fluff and quills of feathers in all sections, pure white.

BUFF-LACED POLISH.

Disqualifications.

Shanks other than blue or slaty-blue. (See general disqualifications.)

COLOR OF MALE.

HEAD: Plumage, rich buff ticked with pale buff.
BEAK: Slaty-blue.
EYES: Reddish-bay.
COMB, FACE AND WATTLES: Bright red.
EAR-LOBES: White.
CREST: Rich buff, each feather laced with pale buff.
BEARD: Rich buff, each feather laced with pale buff.
NECK: Rich buff, each feather laced with pale buff; plumage in front of neck, same as breast.
WINGS: Bows, rich buff, laced with pale buff; coverts, rich buff, each feather laced with pale buff, lacing widest at ends, forming two well defined wing-bars; primaries, buff, the outer end having an edging of pale buff; secondaries, rich buff, with a well defined pale buff lacing.
BACK: Rich buff, each feather laced with pale buff; saddle, rich buff, each feather laced with pale buff.
TAIL: Buff; sickles and coverts, buff, each feather laced with pale buff.
BREAST: Rich buff, each feather laced with pale buff.
BODY AND FLUFF: Body, rich buff, each feather laced with pale buff; fluff, light buff.

LEGS AND TOES: Thighs, buff, laced with pale buff; shanks and toes, slaty-blue.

UNDER-COLOR OF ALL SECTIONS: Pale buff.

COLOR OF FEMALE.

HEAD: Plumage, rich buff ticked with pale buff.

BEAK: Slaty-blue.

EYES: Reddish-bay.

COMB, FACE AND WATTLES: Bright red.

EAR-LOBES: White.

CREST: Buff, each feather laced with pale buff.

BEARD: Buff, each feather laced with pale buff.

NECK: Buff, each feather laced with pale buff; feathers in front of neck, same as breast.

WINGS: Bows, buff, each feather laced with pale buff; coverts, buff, laced with pale buff, pale buff growing wider at extremity, forming two well defined wing-bars; primaries, buff; secondaries, buff, with a well defined lacing of pale buff.

BACK: Buff, each feather laced with pale buff.

TAIL: Buff, each feather laced with pale buff, the pale buff being wider at outer end of feather.

BREAST: Buff, each feather laced with pale buff.

BODY AND FLUFF: Body, buff, each feather laced with pale buff; fluff, light buff.

LEGS AND TOES: Thighs, buff, each feather laced with pale buff; shanks and toes, slaty-blue.

UNDER-COLOR OF ALL SECTIONS: Pale buff.

NON-BEARDED POLISH.

(Golden, Silver and White.)

Same as the bearded varieties in every respect, including disqualifications, shape and color, except that they have no beard. (See general disqualifications.)

CLASS VI.

HAMBURGS.

Breed		*Varieties*
		Golden Spangled
		Silver Spangled
HAMBURGS	{	Golden Penciled
		Silver Penciled
		White
		Black

SCALE OF POINTS FOR THE HAMBURG CLASS.

Symmetry	4
Size	4
Condition	4
Comb	10
Head—Shape 2, Color 2.................................	4
Beak—Shape 2, Color 2.................................	4
Eyes—Shape 2, Color 2.................................	4
Wattles and Ear-Lobes—Shape 5, Color 5................	10
Neck—Shape 4, Color 4.................................	8
Wings—Shape 4, Color 6................................	10
Back—Shape 5, Color 5.................................	10
Tail—Shape 4, Color 6.................................	10
Breast—Shape 4, Color 4...............................	8
Body and Fluff—Shape 3, Color 3.......................	6
Legs and Toes—Shape 2, Color 2...	4

100

HAMBURGS.

Hamburgs originated in Holland and derived their name from the city of Hamburg. They are one of our oldest standard breeds and blood from at least two of the varieties has been used in establishing some of our most popular American breeds. Hamburgs breed remarkably true to type and specimens of all ages show symmetrical outlines. They are rather small; their plumage is close fitting and the markings in the Silver and Golden Spangled and Silver and Golden Penciled varieties rival the pheasant in beauty. The combs of all varieties are rather large for the size of the specimens. The difficulty of securing perfection in the wide range of color found in the several varieties calls forth the greatest skill of the breeders.

Hamburg Disqualifications.

Red in ear-lobes covering more than one-third of the surface; hen feathered male; shanks other than leaden-blue.

SHAPE OF MALE.

HEAD: Short, medium in size.

BEAK: Of medium length; well curved.

EYES: Rather large, round, full.

COMB: Rose; medium in size, not so large as to overhang the eyes or beak; square in front; firm and even on head; uniform on sides; free from hollow center; top covered with small points; fine in texture; terminating at rear in a spike which inclines upward very slightly.

WATTLES AND EAR-LOBES: Wattles, medium in size, thin, well rounded, free from wrinkles. Ear-lobes, of medium size, flat, round, smooth, even, fitting closely to the head.

NECK: Tapering, with full hackle flowing well over shoulders.

WINGS: Large, carried rather low.

BACK: Of medium length, flat at shoulders, straight, gradually sloping to rear of saddle.

TAIL: Full, well expanded, carried at an angle of forty degrees above the horizontal (see illustration, figure 39); sickles, well curved; coverts, abundant.

BREAST: Round, prominent, carried well forward.

BODY AND FLUFF: Body, round, smooth; fluff, rather short.

LEGS AND TOES: Thighs, of medium size and length; shanks, of medium length; toes, straight.

SHAPE OF FEMALE.

HEAD: Short, medium in size.

BEAK: Of medium length; well curved.

EYES: Rather large, round, full.

COMB: Rose; similar to that of male, but smaller.

WATTLES AND EAR-LOBES: Wattles, small, thin, well rounded. Ear-lobes, small, flat, round, smooth, even, fitting closely to head.

NECK: Full at base, tapering to head, slightly arched.

WINGS: Large, carried rather low.

BACK: Of medium length, moderately full, with slight concave sweep to tail.

TAIL: Full, somewhat expanded, carried at an angle of forty degrees above the horizontal. (See illustration, figure 39.)

BREAST: Round, prominent, carried well forward.

BODY AND FLUFF: Body, round, smooth; fluff, rather short.

LEGS AND TOES: Thighs, of medium size, well developed; shanks, rather short, slender; toes, straight.

GOLDEN SPANGLED HAMBURGS.

Disqualifications.

Absence of distinct bars across the wings; markings wholly crescentic. (See general and Hamburg disqualifications.)

COLOR OF MALE.

HEAD: Plumage, golden-bay.

BEAK: Dark horn.

EYES: Reddish-bay.

COMB, FACE AND WATTLES: Bright red.

EAR-LOBES: White.

NECK: Golden-bay, with a glossy, greenish-black stripe extending down middle of each feather, terminating in a point near its lower extremity; plumage in front of neck, same as breast.

WINGS: Fows, rich golden-bay, distinctly spangled with lustrous greenish-black; coverts, golden-bay, free from lacing, each feather ending with a large, greenish-black spangle, forming two distinct parallel bars across wings; primaries, upper webs black, lower webs bay; secondaries, rich golden-bay, lower feathers ending with lustrous greenish-black, crescent-shaped spangles, gradually increasing into V-shaped spangles as they approach the back.

BACK: Lustrous golden-bay, spangled with greenish-black, the texture of the feathers giving the spangles a rayed appearance; saddle feathers, lustrous golden-bay, each feather ending with an elongated greenish-black spangle, the spangle being proportionate to size of feather.

TAIL: Lustrous, greenish-black; sickles, lustrous greenish-black; coverts, lustrous greenish-black.

BREAST: Golden-bay, each feather ending with a large, greenish-black spangle, the spangle being proportionate to size of feather.

BODY AND FLUFF: Body, golden-bay, each feather ending with a large, greenish-black spangle, the spangle being proportionate to size of feather; fluff, bay tinged with slate.

LEGS AND TOES: Thighs, golden-bay, each feather ending with a greenish-black spangle; shanks and toes, leaden-blue.

UNDER-COLOR OF ALL SECTIONS: Slate.

COLOR OF FEMALE.

HEAD: Plumage, rich, golden-bay.

BEAK: Dark horn.

EYES: Reddish-bay.

COMB, FACE AND WATTLES: Bright red.

EAR-LOBES: White.

NECK: Golden-bay, with a lustrous greenish-black stripe extending down middle of each feather, terminating in a point near its lower extremity; feathers in front of neck, same as breast.

WINGS: Bows, golden-bay, distinctly spangled with lustrous greenish-black; coverts, clear, reddish-bay, free from lacing, each feather ending with a large, greenish-black spangle, forming two distinct parallel bars across wings; primaries, upper webs black, lower webs bay; secondaries, golden bay, each feather ending with a lustrous greenish-black, crescent-shaped spangle, gradually increasing into V-shaped spangle as it approaches the back.

BACK: Golden-bay, each feather ending with a large, greenish-black spangle, the spangle being proportionate to size of feather.

TAIL: Greenish-black; coverts, golden-bay, each feather ending with a lustrous greenish-black spangle.

BREAST: Golden-bay, each feather ending with a large, greenish-black spangle, the spangle being proportionate to size of feather.

BODY AND FLUFF: Body, golden-bay, each feather ending with a large, greenish-black spangle, the spangle being proportionate to size of feather; fluff, slate tinged with gray.

LEGS AND TOES: Thighs, golden-bay, each feather ending with a greenish-black spangle; shanks and toes, leaden-blue.

UNDER-COLOR OF ALL SECTIONS: Slate.

Note: In all sections where the word "spangle" appears, when shape is not otherwise described, read "edges of spangle following web of feather and meeting at shaft."

SILVER SPANGLED HAMBURGS.

Disqualifications.

Absence of distinct bars across wings; markings wholly crescentic. (See general and Hamburg disqualifications.)

COLOR OF MALE.

HEAD: Plumage, white.

BEAK: Dark horn.

EYES: Reddish-bay.

COMB, FACE AND WATTLES: Bright red.

EAR-LOBES: White.

NECK: White, each feather ending with an elongated black spangle, the spangle being proportionate to size of feather; plumage in front of neck, same as breast.

WINGS: Bows, silvery white, distinctly spangled with black; coverts, silvery white, free from lacing, each feather ending with a large, black spangle, forming two distinct parallel bars across wings; primaries, white, each feather edged with black at end; secondaries, white, each feather ending with a black, crescent-shaped spangle, gradually increasing into a V-shaped spangle as it approaches the back.

BACK: Clear, silvery white, each feather ending with an elongated black spangle, the texture of feathers giving spangles a rayed appearance; saddle, clear, silvery white, each feather ending with an elongated black spangle, the spangle being proportionate to size of feather.

TAIL: White, each feather ending with a large, black spangle;

sickles, pure white, ending with a large, black spangle; coverts, pure white, ending with black spangles.

BREAST: Clear, silvery-white, each feather ending with a large, black spangle, the spangle being proportionate to size of feather.

BODY AND FLUFF: Body, silvery-white, each feather ending with a large, black spangle, the spangle being proportionate to size of feather; fluff, slate, tinged with white.

LEGS AND TOES: Thighs, silvery-white, each feather ending with a black spangle; shanks and toes, leaden-blue.

UNDER-COLOR OF ALL SECTIONS: Slate.

COLOR OF FEMALE.

HEAD: Plumage, white.

BEAK: Dark horn.

EYES: Reddish-bay.

COMB, FACE AND WATTLES: Bright red.

EAR-LOBES: White.

NECK: Silvery-white, each feather ending with an elongated, small, black spangle; feathers in front of neck, same as breast.

WINGS: Bows, silvery-white, distinctly spangled with black; coverts, silvery-white, each feather ending with a large, black spangle, forming two distinct parallel bars across wings; primaries, white, each feather ending with black; secondaries, white, each feather ending with a lustrous black, crescent-shaped spangle, gradually increasing into a V-shaped spangle, as it approaches the back.

BACK: Silvery-white, each feather ending with a large, black spangle.

TAIL: White, each feather ending with a large, black spangle; coverts, white on the outside, each feather ending with a black spangle.

BREAST: Clear, silvery-white, each feather ending with a black spangle, the spangle being proportionate to size of feather.

BODY AND FLUFF: Body, silvery-white, each feather ending with a black spangle, the spangle being proportionate to size of feather; fluff, slate, tinged with white.

LEGS AND TOES: Thighs, silvery-white, each feather ending with a black spangle; shanks and toes, leaden-blue.

UNDER-COLOR OF ALL SECTIONS: Slate.

Note: In all sections where the word "spangle" appears when shape is not otherwise described, read "edges of spangle following the web of feather and meeting at shaft."

SILVER SPANGLED HAMBURG MALE

SILVER SPANGLED HAMBURG FEMALE

GOLDEN PENCILED HAMBURGS.

Disqualifications.

Breast of female not penciled. (See general and Hamburg disqualifications.)

COLOR OF MALE.

HEAD: Plumage, rich, bright bay.

BEAK: Dark horn.

EYES: Reddish-bay.

COMB, FACE AND WATTLES: Bright red.

EAR-LOBES: White.

NECK: Rich, bright reddish-bay; plumage in front of neck, penciled same as breast.

WINGS: Bows, bright, reddish-bay; coverts, reddish-bay, upper web slightly penciled across with black bars; primaries, upper webs black, lower webs bay; secondaries, upper webs reddish-bay, penciled across with black bars, lower webs, reddish bay, each feather ending with a small, black spot.

BACK: Rich, bright reddish-bay.

TAIL: Black; sickles and coverts, greenish-black with a distinct edging of rich reddish-bay, the narrower the better.

BREAST: Rich reddish-bay.

BODY AND FLUFF: Body, lustrous reddish-bay, the sides below wings penciled across with indistinct black bars; fluff, black.

LEGS AND TOES: Thighs, reddish-bay; shanks and toes, leaden-blue.

UNDER COLOR OF ALL SECTIONS: Slate.

COLOR OF FEMALE.

HEAD: Plumage, bright reddish-bay.

BEAK: Dark horn.

EYES: Reddish-bay.

COMB, FACE AND WATTLES: Bright red.

EAR-LOBES: White.

NECK: Bright bay; feathers in front of neck, same as breast.

WINGS: Bows, clear bay, finely and distinctly penciled across with greenish-black; primaries, bay; secondaries and coverts, bay, penciled across with greenish-black.

BACK: Bay, each feather finely and distinctly penciled across with narrow parallel bars of greenish-black.

TAIL: Bay, penciled across with greenish-black; coverts, bay, penciled across with greenish-black.

234

BREAST: Bright bay, each feather finely and distinctly penciled across with parallel bars of greenish-black.

BODY AND FLUFF: Body, bay, each feather finely and distinctly penciled across with parallel bars of greenish-black, the bars forming, as nearly as possible, narrow, parallel lines across the specimen; fluff, bay, penciled with black.

LEGS AND TOES: Thighs, bay, penciled across with greenish-black; shanks and toes, leaden-blue.

UNDER COLOR OF ALL SECTIONS: Slate.

SILVER PENCILED HAMBURGS.

Disqualifications.

Breast of female not penciled. (See general and Hamburg disqualifications.)

COLOR OF MALE.

HEAD: Plumage, white.

BEAK: Dark horn.

EYES: Reddish-bay.

COMB, FACE AND WATTLES: Bright red.

EAR-LOBES: White.

NECK: Clear white; plumage in front of neck, same as breast.

WINGS: Bows, white; coverts, apparently white when wings are folded, but penciled with black on upper webs; primaries, white; secondaries, upper webs black with a narrow border of white or gray on edges; lower webs, white, with a narrow stripe of black next to shafts of feathers.

BACK: Silvery white; saddle feathers, silvery white.

TAIL: Black; sickles and coverts, black with distinct edging of white, the narrower and more uniform, the better.

BREAST: White.

BODY AND FLUFF: Body, white, the sides below wings penciled across with indistinct black bars; fluff, slaty-white.

LEGS AND TOES: Thighs, silvery white; shanks and toes, leaden-blue.

UNDER COLOR OF ALL SECTIONS: Slate.

COLOR OF FEMALE.

HEAD: Plumage, white.

BEAK: Dark horn.

EYES: Reddish-gray.

COMB, FACE AND WATTLES: Bright red.

EAR-LOBES: White.

NECK: White, except at base, which should be penciled across with narrow bars of greenish-black; feathers in front of neck, same as breast.

WINGS: Bows, white, finely and distinctly penciled across with greenish-black; primaries, white; secondaries and coverts, white, penciled across with greenish-black.

BACK: White, each feather finely and distinctly penciled across with narrow, parallel bars of greenish-black.

TAIL: White, penciled across with greenish-black; coverts, silvery white, penciled across with greenish-black.

BREAST: White, each feather finely and distinctly penciled across with narrow, parallel bars of greenish-black.

BODY AND FLUFF: Body, white, each feather finely and distinctly penciled across with narrow, parallel bars of greenish-black, the bars forming as nearly as possible, narrow, parallel lines across the specimen; fluff, white, penciled with black.

LEGS AND TOES: Thigh, white, penciled across with greenish-black; shanks and toes, leaden-blue.

UNDER COLOR OF ALL SECTIONS: Slate.

WHITE HAMBURGS.

Disqualifications.

Feathers other than white in any part of plumage. (See general and Hamburg disqualifications.)

COLOR OF MALE AND FEMALE.

BEAK: Leaden blue.
EYES: Reddish-bay.
COMB, FACE AND WATTLES: Bright red.
EAR-LOBES: White.
SHANKS AND TOES: Leaden-blue.
PLUMAGE: Web, fluff and quills of feathers in all sections, pure white.

BLACK HAMBURGS.

Disqualifications.

White in the face of cockerels or pullets; pure white in any part of plumage extending over half an inch, or two or more feathers tipped or edged with positive white; shanks other than leaden-blue or black. (See general and Hamburg disqualifications.)

COLOR OF MALE AND FEMALE.

BEAK: Black.
EYES: Reddish-bay.
COMB, FACE AND WATTLES: Bright red.
EAR-LOBES: White.
SHANKS AND TOES: Black.
PLUMAGE: Surface, lustrous greenish-black throughout. Undercolor of all sections, dull black.

CLASS VII.

FRENCH.

Breeds	Varieties
HOUDANS..	{ Mottled
	{ White
CREVECOEURS....................................	Black
LA FLECHE......................................	Black
FAVEROLLES.....................................	Salmon

SCALE OF POINTS FOR THE FRENCH CLASS.

(Except for La Fleche which are subject to Scale of points for American Class.)

Symmetry	4
Weight	4
Condition	4
Comb	4
Head—Shape 2, Color 2.................................	4
Beak—Shape 2, Color 2.................................	4
Eyes—Shape, 2, Color 2................................	4
Wattles and Ear-Lobes—Shape 2, Color 2.................	4
Neck—Shape 4, Color 4.................................	8
Wings—Shape 4, Color 6...............................	10
Back—Shape 5, Color 4................................	9
Tail—Shape 5, Color 4................................	9
Breast—Shape 6, Color 4..............................	10
Body and Fluff—Shape 3, Color 3.......................	6
Legs and Toes—Shape 2, Color 2.......................	4
Crest and Beard—Shape 8, Color 4.....................	12
	100

Omit crest in Faverolles and apply the 12 points allowed for this section to Beard and Muffs.

HOUDANS.

The Houdan is one of the oldest and best known of the French breeds. In shape and size it is similar to the Dorkings, and like it, the Houdan is highly esteemed for its meat qualities, while those who breed these birds greatly admire their glossy black feathers, regularly tipped with white and the full, round crest which is one of the most distinctive characteristics of the breed. The Standard weights are not placed unduly high, hence their attainment should be common in well bred fowls.

Disqualifications.

Absence of crest or beard; feathers other than black or white in any part of plumage. (See general disqualifications.)

STANDARD WEIGHTS.

Cock ....7½ lbs.   Hen ....6½ lbs.
Cockerel ....6½ lbs.   Pullet ....5½ lbs.

SHAPE OF MALE.

HEAD: Of medium size, carried well up.
BEAK: Of moderate length, well curved.
NOSTRILS: Wide, cavernous.
EYES: Large, full.
COMB AND CREST: Comb, V-shaped; of small size, resting against front of crest. Crest, large, well fitted on crown of head, falling backward on neck, and composed of feathers similar in shape and texture to those of hackle.
BEARD: Full, well developed, curving around to back of eyes, nearly hiding face.
WATTLES AND EAR-LOBES: Wattles, of uniform length, small, well-rounded, nearly concealed by beard. Ear-lobes, entirely concealed by crest and beard.
NECK: Of medium length, well arched; with abundant hackle flowing well down on shoulders.
WINGS: Moderately large, well folded, fronts concealed by breast feathers and points by saddle feathers.
BACK: Long, broad, slightly sloping toward base of tail; saddle feathers, abundant.

TAIL: Full, expanded, carried at an angle of forty degrees above the horizontal (see illustration, figure 39); sickles and coverts, abundant and well curved.

BREAST: Broad, deep, full and well rounded.

BODY AND FLUFF: Body, long, compact, well proportioned; fluff, rather short.

LEGS AND TOES: Thighs, of medium length, large, set well apart; shanks, of medium length; toes, five on each foot, straight, except the fifth, which should be detached from the others and curve upwards.

SHAPE OF FEMALE.

HEAD: Of medium size.

BEAK: Of moderate length, well curved.

NOSTRILS: Wide, cavernous.

EYES: Large, full.

COMB AND CREST: Comb, V-shaped, similar to that of male. Crest, large, compact, regular, inclining backward in an unbroken mass.

BEARD: Full, well developed, curving around to back of eyes, nearly hiding face.

WATTLES AND EAR-LOBES: Wattles, of uniform length, small, well-rounded, nearly concealed by beard. Ear-lobes, entirely concealed by crest and beard.

NECK: Of medium length, well arched.

WINGS: Moderately large, well folded.

BACK: Long, broad, slightly sloping toward base of tail.

TAIL: Of medium length, rather compact; carried at an angle of forty degrees above the horizontal. (See illustration, figure 39.)

BREAST: Broad, deep, full, well rounded.

BODY AND FLUFF: Body, long, compact, well proportioned; fluff, rather short.

LEGS AND TOES: Thighs, short, strong, set well apart, shanks of medium length; toes, five on each foot, straight, except fifth, which should be detached from the others and curve slightly upward.

MOTTLED HOUDANS.

COLOR OF MALE.

HEAD: Plumage, black and white, black predominating.
BEAK: Dark horn.
EYES: Reddish-bay.
COMB, FACE AND WATTLES: Bright red.
EAR-LOBES: White.
CREST AND BEARD: Black and white, black predominating.
NECK: Black, about one feather in five tipped with positive white.
WINGS: Bows and coverts, black, about one feather in five tipped with positive white; primaries, black and white, colors pure within themselves, black predominating; secondaries, black.
BACK: Black, a slight ticking of white, about one feather in ten, not a serious defect.
TAIL: Black, ends of feathers, in proportion of about one in four, tipped with positive white; sickle feathers, black, which may be edged with white; coverts, black, about one feather in five tipped with positive white.
BREAST: Black, about one feather in five tipped with positive white.
BODY AND FLUFF: Body, black, about one feather in five tipped with positive white; fluff, black, tipped with gray.
LEGS AND TOES: Thighs, black, about one feather in five tipped with positive white; shanks and toes, pinkish-white, mottled with black.
UNDER-COLOR OF ALL SECTIONS: Dull black.

COLOR OF FEMALE.

HEAD: Plumage, black and white, black predominating.
BEAK: Dark horn.
EYES: Reddish-bay.
COMB, FACE AND WATTLES: Bright red.
EAR-LOBES: White.
CREST AND BEARD: Black and white, black predominating.
NECK: Black, about one feather in five tipped with positive white.
WINGS: Bows and coverts, black, about one feather in five tipped with positive white; primaries, black and white, colors pure within themselves, black predominating; secondaries, black.

HOUDAN MALE

242

HOUDAN FEMALE

BACK: Black, about one feather in five tipped with positive white.

TAIL: Black, ends of feathers, in proportion of about one to four, tipped with positive white; coverts, black, about one feather in five tipped with positive white.

BREAST: Black, about one feather in five tipped with positive white.

BODY AND FLUFF: Body, black, about one feather in five tipped with positive white; fluff, black tipped with gray.

LEGS AND TOES: Thighs, black, about one feather in five tipped with positive white; shanks and toes, pinkish-white, mottled with black.

UNDER-COLOR OF ALL SECTIONS: Dull black.

WHITE HOUDANS.

Disqualifications.

Absence of crest or beard; feathers other than white in any part of plumage. (See general disqualifications.)

BEAK: Pinkish-white.

EYES: Reddish-bay.

COMB, FACE, WATTLES AND EAR-LOBES: Red.

SHANKS AND TOES: Pinkish-white.

PLUMAGE: Web, fluff and quill of feathers in all sections, pure white.

CREVECOEURS.

The Crevecoeurs is one of the oldest of the French breeds and takes its name from the village of Crevecoeur. The breed has not taken a strong hold on the American fancy and comparatively few are to be found in this country. They are handsome and useful fowls. The plumage color is a solid, glossy black throughout; crest, well set on crown of head, and composed of feathers similar in shape and texture to those of neck.

Disqualifications.

Absence of crest or beard; pure white in any part of plumage extending over half an inch, or two or more feathers tipped or edged with positive white, except in crest; shanks other than black or dark lead color. (See general disqualifications.)

STANDARD WEIGHTS.

Cock	.....8 lbs.	Hen	.....7 lbs.
Cockerel	.....7 lbs.	Pullet	.....6 lbs.

SHAPE OF MALE.

HEAD: Large with a pronounced protuberance on top of skull.

FACE: Almost wholly concealed by crest and beard.

BEAK: Strong, well curved.

NOSTRILS: Broad, highly arched.

EYES: Full, oval.

COMB AND CREST: Comb, like the letter V in shape, of medium size, resting against front of crest. Crest, large, well fitted on crown of head, regular, inclining backward, composed of feathers similar in shape and texture to those of hackle.

BEARD: Full, thick, extending around to back of eyes, nearly hiding face.

WATTLES AND EAR-LOBES: Wattles, of uniform length, small, well rounded, nearly concealed by beard. Ear-lobes, small, nearly concealed by crest and beard.

NECK: Of medium length, well arched, with abundant hackle, flowing well down on shoulders.

WINGS: Of medium size, well folded.

BACK: Broad, straight; saddle feathers abundant.

245

TAIL: Full, expanded, carried at an angle of forty-five degrees above the horizontal (see illustration, fig. 39), abundant, well curved.

BREAST: Broad, full, rounding well to shoulders.

BODY AND FLUFF: Body, compact, well proportioned; fluff, rather short.

LEGS AND TOES: Thighs, short, strong; shanks, short, fine in bone, standing well apart; toes, four on each foot, straight, well spread.

SHAPE OF FEMALE.

HEAD: Large, with a pronounced protuberance on top of skull.

FACE: Almost wholly concealed by crest and beard.

BEAK: Strong, well curved.

NOSTRILS: Broad, highly arched.

EYES: Full, oval.

COMB AND CREST: Comb, like letter V in shape, small and as nearly as possible concealed by crest. Crest, large, compact, even, globular, inclining backward in an unbroken mass.

BEARD: Full, thick, extending around to back of eyes, nearly hiding face.

WATTLES AND EAR-LOBES: Wattles, of uniform length, small, well rounded, nearly concealed by beard. Ear-lobes, small, entirely concealed by crest and beard.

NECK: Of medium length, thick, well arched.

WINGS: Of medium size, well folded.

BACK: Broad, straight.

TAIL: Moderately expanded at base, converging to tip, carried at an angle of forty-five degrees above the horizontal. (See illustration, fig. 39.)

BREAST: Broad, full, rounding well to shoulders.

BODY AND FLUFF: Body, compact, well proportioned; fluff, rather short.

LEGS AND TOES: Thighs, short, strong; shanks, rather short, medium in bone; toes, four on each foot, straight, well spread.

COLOR OF MALE AND FEMALE.

BEAK: Black, shading into horn at tip.

EYES: Bright red.

COMB, FACE, WATTLES AND EAR-LOBES: Red.

SHANKS AND TOES: Black or dark leaden.

PLUMAGE: Rich, glossy black.

UNDER-COLOR OF ALL SECTIONS: Dull black.

LA FLECHE.

The La Fleche are of French origin and supposed to be from a cross of the Crevecoeurs and Spanish. With their long and powerful bodies and solid black plumage they present a striking and somewhat fierce appearance.

Disqualifications.

Presence of crest; pure white in any part of plumage extending over half an inch, or two or more feathers edged or tipped with positive white; shanks other than black or slate in color.

STANDARD WEIGHTS.

Cock ....8½ lbs.   Hen ....7½ lbs.
Cockerel ....7½ lbs.   Pullet ....6½ lbs.

SHAPE OF MALE.

HEAD: Of medium size, long.

BEAK: Rather long, strong, well curved.

NOSTRILS: Wide, cavernous.

EYES: Large.

COMB: V-shaped; rather large.

WATTLES AND EAR-LOBES: Wattles, of equal length, long, well rounded, pendulous. Ear-lobes, large.

NECK: Long, erect, with abundant hackle flowing well down on shoulders.

WINGS: Long, powerful, well folded.

BACK: Broad, very long, slanting to tail; saddle feathers, abundant.

TAIL: Very long, full, carried at an angle of forty degrees above the horizontal. (See illustration, figure 39.)

BREAST: Broad, full, very prominent.

BODY AND FLUFF: Body, large, powerful, tapering to tail, with close plumage; fluff, rather short.

LEGS AND TOES: Thighs, long, powerful; shanks long, medium in bone; toes, straight, large.

SHAPE OF FEMALE.

HEAD: Of medium size, long.

BEAK: Rather long, strong, well curved.

NOSTRILS: Wide, cavernous.

EYES: Large.

COMB: V-shaped, medium large.

WATTLES AND EAR-LOBES: Wattles, of equal length, small, well rounded. Ear-lobes, small.

NECK: Long, carried upright, with full plumage.

WINGS: Long, powerful, well folded.

BACK: Broad, long, slanting to tail.

TAIL: Long, well expanded, carried at an angle of sixty degrees above the horizontal. (See illustration, figure 39.)

BREAST: Broad, full, prominent.

BODY AND FLUFF: Body, large, deep, tapering to tail; fluff, rather short.

LEGS AND TOES: Thighs, long, powerful; shanks long, medium in bone; toes, straight, large.

COLOR OF MALE AND FEMALE.

BEAK: Black, or dark horn, with a small protuberance of bright red flesh at juncture of beak and nostrils.

EYES: Bright red.

COMB, FACE AND WATTLES: Bright red.

EAR-LOBES: White.

SHANKS AND TOES: Black or slate.

PLUMAGE: Rich, glossy black.

UNDER-COLOR OF ALL SECTIONS: Dull black.

FAVEROLLES.

Faverolles originated in Faverolle, France. The origin was a cross of the Houdans, Dorkings and Asiatics. They possess a combination of unique qualities; the beard and muffs, the variety of colors, the full breast, square back and deep body, all of which offer problems of absorbing interest for the breeder to solve.

STANDARD WEIGHTS.

Cock ....8 lbs.   Hen ....6½ lbs.
Cockerel ....7 lbs.   Pullet ....5½ lbs.

Disqualifications.

Absence of beard or muffs. (See general disqualifications.)

SHAPE OF MALE.

HEAD: Broad, flat, short, free from crest.
BEAK: Short, stout, well curved.
EYES: Large, full.
COMB: Single; of moderate size, straight and upright, evenly serrated, having five well-defined points, the front and rear shorter than the other three, fine in texture.
BEARD AND MUFFS: Full, wide, short.
WATTLES AND EAR-LOBES: Wattles, small, well rounded, fine in texture. Ear-lobes, oblong, concealed by beard.
NECK: Short, thick, well arched.
WINGS: Moderately small, prominent in front, carried closely folded to body.
BACK: Broad, flat, almost square.
TAIL: Of moderate length, carried at an angle of fifty degrees above the horizontal (see illustrations, figures 39 and 41); sickles and coverts, of moderate length and well curved.
BREAST: Broad, deep, carried well forward.
BODY AND FLUFF: Body, deep, compact; fluff, rather short.
LEGS AND TOES: Thighs, moderate length, stout, straight; set well apart; toes, five on each foot, the fifth toe separate from the others and curved upward; outer toes, slightly feathered.

249

SHAPE OF FEMALE.

HEAD: Rather broad, flat, short, free from crest.

BEAK: Short, stout, well curved.

EYES: Large, full.

COMB: Single; of moderate size, straight and upright, evenly serrated, having five well defined points, the front and rear shorter than the other three, fine in texture.

BEARD AND MUFFS: Full, wide, short.

WATTLES AND EAR-LOBES: Wattles, small, well rounded, fine in texture. Ear-lobes, oblong, concealed by beard.

NECK: Short, thick, fairly well arched.

WINGS: Moderately small, carried closely folded to body.

BACK: Broad, flat, longer than wide.

TAIL: Full, moderate length, carried at an angle of fifty degrees above the horizontal. (See illustration, figure 39.)

BREAST: Broad, deep, prominent.

BODY AND FLUFF: Body, very deep, long; fluff, rather short.

LEGS AND TOES: Thighs, moderate length, rather stout, straight, set well apart; toes, five on each foot, the fifth toe separate from the others and curved upward; outer toes slightly feathered.

SALMON FAVEROLLES.

COLOR OF MALE.

HEAD: Plumage, straw.

BEAK: Horn.

EYES: Reddish-bay.

COMB, FACE, EAR-LOBES AND WATTLES: Bright red.

BEARD AND MUFFS: Black.

NECK: Hackle, straw; plumage in front of hackle, same as breast.

WINGS: Bows, straw; bars, black, primaries, black, lower elge white; secondaries, black, lower one-third of outer web white.

BACK: Web, the outer portion reddish-brown, edged with a lighter shade of brown, portion nearest to undercolor, black; saddle, straw.

TAIL: Main tail, black; sickles, smaller sickles and coverts, greenish-black.

BREAST: Black.

BODY AND FLUFF: Black.

LEGS AND TOES: Thighs, black; shanks and toes white or pink-ish-white.

UNDER-COLOR OF ALL SECTIONS: Slate.

COLOR OF FEMALE.

HEAD: Plumage, salmon-brown.

BEAK: Horn.

EYES: Reddish-bay.

COMB, FACE, EAR-LOBES AND WATTLES: Bright red.

BEARD AND MUFFS: Creamy white.

NECK: Rich salmon-brown; feathers in front of neck, same as breast.

WINGS: Bows and covert, rich salmon-brown; primaries, the lower web slightly edged with salmon-brown, upper web slightly stippled with salmon-brown; secondaries, lower half of outer web, salmon-brown, upper half, black.

BACK: Salmon-brown..

TAIL: Salmon-brown.

BREAST: Cream.

BODY AND FLUFF: Cream.

LEGS AND TOES: Thighs, cream; shanks and toes, white or pinkish-white.

UNDER-COLOR OF ALL SECTIONS: Slate.

CONTINENTAL.

Breeds	Varieties
CAMPINES..	{ Silver { Golden

SCALE OF POINTS.

Symmetry	4
Weight	4
Condition	4
Comb	8
Head—Shape 2, Color 3...................................	5
Eyes—Shape 2, Color 2..................................	4
Beak—Shape 2, Color 2..................................	4
Wattles and Ear-Lobes—Shape 4, Color 4.................	8
Neck—Shape 3, Color 5..................................	8
Wings—Shape 4, Color 6.................................	10
Back—Shape 6, Color 6..................................	12
Tail—Shape 5, Color 5..................................	10
Breast—Shape 4, Color 5................................	9
Body and Fluff—Shape 3, Color 3,.......................	6
Legs and Toes—Shape 2, Color 2........................	4
	100

CAMPINES.

This breed was originated from two similar breeds of fowls, namely the Campines and the Braekel of Belgium. The present Campine is much improved in markings over both of the original breeds. Both males and females are alert. This activity, along with their handsome markings, makes them especially attractive. The Campines should be deep, long-bodied, not too broad across the shoulders, and rather flat on backs. Short backs and short shanks should be avoided. The plumage being close fitting, the specimens carry more weight than appearance indicates. The two varieties should be identical except in color.

Disqualifications.

Red covering more than one-half of the ear-lobes; white in the face of cockerels; legs other than leaden-blue. (See general disqualifications.)

STANDARD WEIGHTS.

Cock ....6 lbs.   Hen ....4 lbs.
Cockerel ....5 lbs.   Pullet ....3½ lbs.

SHAPE OF MALE.

HEAD: Of medium length, fairly deep; face, smooth, fine in texture.

BEAK: Of medium length, nicely curved.

EYES: Large, nearly round, prominent.

COMB: Single, medium size, straight, upright, firm and even on head, having five distinct points, deeply serrated; extending well over back of head; blade carried slightly below the horizontal; smooth, free from twists, folds and excrescences.

WATTLES AND EAR-LOBES: Wattles, long, thin, even in length, well rounded; smooth in texture; free from folds or wrinkles. Ear-Lobes, oval in shape but rather broad, smooth, of moderate size, fitting closely to head.

NECK: Medium length, nicely arched, and well furnished with hackle feathers.

WINGS: Large, well folded and tucked up.

BACK: Rather long, slightly sloping to tail, not too broad at shoulders and narrowing very slightly toward tail; back feathers, abundant, long, wide, ending with rounded tips.

TAIL: Well expanded, main feathers carried at an angle of forty-five degrees above the horizontal (see illustration, figure 39); sickles well curved and extending beyond main tail feathers; smaller sickles and coverts, the more abundant, the better.

BREAST: Deep, well rounded, and carried well forward.

BODY AND FLUFF: Body of moderate length, and fairly deep; not narrow in appearance from behind; fluff, moderately short.

LEGS AND TOES: Thighs and shanks, rather long and slender; shanks, round; toes, rather long, slender and straight.

SHAPE OF FEMALE.

HEAD: Moderate in length, fairly deep, well rounded; face, smooth, fine in texture.

BEAK: Moderate in length, nicely curved.

EYES: Large, nearly round, prominent.

COMBS Single; medium in size, deeply serrated, having five distinct points; the front portion of comb and first point to stand erect, and the remainder of comb drooping gradually to one side; fine in texture, free from folds or wrinkles.

WATTLES AND EAR-LOBES: Wattles, moderate size, well rounded, fine texture. Ear-lobes, oval in shape, smooth, thin, fitting closely to the head.

NECK: Of medium length, slender, slightly arched.

WINGS: Large, well folded and tucked up.

BACK: Rather long, declining slightly to tail, not too broad at shoulders and narrowing very slightly toward tail, somewhat rounded across cape.

TAIL: Long, full, moderately spread, carried at an angle of forty degrees above the horizontal. (See illustration, figure 39.)

BREAST: Deep, well rounded and carried well forward.

BODY AND FLUFF: Body of moderate length, fairly deep; not narrow in appearance from behind; fluff, moderately short.

LEGS AND TOES: Thighs and shanks rather long and slender; shanks, round; toes, rather long, slender and straight.

SILVER CAMPINES.

COLOR OF MALE.

HEAD: Plumage, white.

BEAK: Horn.

EYES: Dark brown.

COMB, FACE, WATTLES: Red.

EAR-LOBES: White.

NECK: Hackle, white; slight barring at extreme base, not a serious defect; plumage in front of hackle, same as breast.

WINGS: Bows, greenish black, distinctly barred with white, the black being four times the width of the white bar which is slightly V-shaped; primaries, black, barred straight across with white; secondaries and coverts, greenish-black, barred straight across with white.

BACK: Greenish-black, distinctly barred with white, the black being four times the width of the white bar which is slightly V-shaped.

TAIL: Greenish-black, barred straight across with white; sickles and coverts, greenish-black, distinctly barred with white, the black being four times the width of the white bar which is slightly V-shaped.

BREAST: Greenish-black, barred straight across with white, the black being three times the width of the white bar.

BODY AND FLUFF: Greenish-black, barred straight across with white, the black being four times the width of the white bar.

SHANKS AND TOES: Leaden-blue.

UNDER-COLOR OF ALL SECTIONS: Slate.

COLOR OF FEMALE.

HEAD: Plumage, white.

BEAK: Horn.

EYES: Dark brown.

COMB, FACE AND WATTLES: Red, some blue at base of comb, not a serious defect.

EAR-LOBES: White.

NECK: White, at extreme base, slight barring not a serious defect; feather in front of neck, same as breast.

WINGS: Bows, greenish-black, distinctly barred with white, the black being four times the width of the white bar, which is slightly V-shaped; primaries, black, barred straight across with

SILVER CAMPINE MALE

SILVER CAMPINE FEMALE

white; secondaries and coverts, greenish-black barred straight across with white.

BACK: Greenish-black, distinctly barred with white, the black being four times the width of the white bar, which is slightly V-shaped.

TAIL: Greenish-black, distinctly barred straight across with white, the black being four times the width of the white bar.

BREAST: Greenish-black, distinctly barred with white, the black being equal in width of the white bar at the throat and increasing to three times the width of the white bar at the body, the barring running straight across the feathers.

BODY AND FLUFF: Greenish-black, distinctly barred with white, the black being four times the width of the white bar and running straight across the feathers.

SHANKS AND TOES: Leaden-blue.

UNDER-COLOR OF ALL SECTIONS: Slate.

GOLDEN CAMPINES.

COLOR OF MALE.

HEAD: Plumage, golden-bay.

BEAK: Horn.

EYES: Dark brown.

COMB, FACE AND WATTLES: Red.

EAR-LOBES: White.

NECK: Hackle, golden-bay, slight barring at extreme base not a serious defect; plumage in front of hackle, same as breast.

WINGS: Bows, greenish-black, distinctly barred with golden-bay, the black being four times the width of the golden-bay bar, which is slightly V-shaped; primaries, black barred straight across with golden-bay; secondaries and coverts, greenish-black barred straight across with golden-bay.

BACK: Greenish-black, distinctly barred with golden-bay, the black being four times the width of the golden-bay bar, which is slightly V-shaped.

TAIL: Greenish-black, barred straight across with golden-bay; sickles and coverts, greenish-black, distinctly barred with golden-bay, the black being three times the width of the golden-bay bar, which is slightly V-shaped.

BREAST: Greenish-black, barred straight across with golden-bay, the black being three times the width of the golden-bay bar.

BODY AND FLUFF: Greenish-black, barred straight across with

golden-bay, the black being four times the width of the golden-bay bar.

SHANKS AND TOES: Leaden-blue.

UNDER-COLOR OF ALL SECTIONS: Slate.

COLOR OF FEMALE.

HEAD: Plumage, golden-bay.

BEAK: Horn.

EYES: Dark brown.

COMB, FACE AND WATTLES: Red, some blue at base of comb, not a serious defect.

EAR-LOBES: White.

NECK: Golden-bay, slight barring at extreme base not a serious defect; feathers in front of neck, same as breast.

WINGS: Bows, greenish-black, distinctly barred with golden-bay, the black being four times the width of the golden-bay bar, which is slightly V-shaped; primaries, black barred straight across with golden-bay; secondaries and coverts, greenish-black barred straight across with golden-bay.

BACK: Greenish-black, distinctly barred with golden-bay, the black being four times the width of the golden-bay bar, which is slightly V-shaped.

TAIL: Greenish-black, distinctly barred straight across with golden-bay, the black being four times the width of the golden-bay bar.

BREAST: Greenish-black, distinctly barred with golden-bay, the black being equal in width of the golden-bay bar at the throat and increasing to three times the width of the golden-bay bar at the body, the barring running straight across the feathers.

BODY AND FLUFF: Greenish-black, distinctly barred with golden-bay, the black being four times the width of the golden-bay bar and running straight across the feather.

LEGS AND TOES: Leaden-blue.

UNDER-COLOR OF ALL SECTIONS: Slate.

GAMES AND GAME BANTAMS.

Breeds	Varieties
GAMES..............................	Black-Breasted Red Brown-Red Golden Duckwing Silver Duckwing Birchen Red Pyle White Black
GAME BANTAMS	Black-Breasted Red Brown-Red Golden Duckwing Silver Duckwing Birchen Red Pyle White Black

SCALE OF POINTS.

Station	10
Condition	6
Comb	2
Head—Shape 4, Color 1..................................	5
Beak	4
Eyes	4
Ear-Lobes and Wattles..................................	2
Neck—Shape 5, Color 3..................................	8
Wings—Shape 4, Color 6.................................	10
Back—Shape 4, Color 3..................................	7
Tail—Shape 5, Color 3..................................	8
Breast—Shape 4, Color 3................................	7
Body and Stern—Shape 4, Color 3........................	7
Legs and Toes—Shape 10, Color 4........................	14
Shortness of Feathers..................................	6
	100

GAMES.

The Game has a style of carriage peculiar to itself, which is generally described by the word "Station." A high-stationed specimen is desired. Shortness and closeness of feathering are of great importance, as loose feathered specimens invariably fail in shape of neck. The comb and wattles of the cock should be dubbed, in order to have the head and lower mandible smooth and free from ridges. Exceptionally large specimens are undesirable, as overgrowth tends to coarseness at the expense of form and style of carriage, which are essential to superior quality in Games.

Disqualifications.

Cocks not dubbed; artificial coloring; trimming or plucking of feathers; duckfoot. (See general disqualifications.)

SHAPE OF MALE.

HEAD: Long, lean, bony.

FACE: Lean, thin, with fine skin.

BEAK: Long, tapering, slightly curved.

EYES: Large, full, with keen expression.

COMB: Cock, neatly and smoothly dubbed; cockerel, if undubbed, single, small, straight, thin, erect, evenly serrated.

WATTLES AND EAR-LOBES: Cock, neatly and smoothly dubbed; cockerel, if undubbed, small, thin, round, smooth.

NECK: Long, very slightly arched, carried erect, tapering neatly and gradually from body to head, thin and clean-cut at throat, giving a distinct outline to head; hackle, short, close.

WINGS: Large, powerful, the front standing out from body at shoulders, the feathers folded closely together, the points not extending beyond body; carried without drooping, but not carried over the back.

BACK: Flat, rather short, straight on top from hackle to tail, broad at shoulders, narrowing and sloping to stern.

TAIL: Rather short, compact, closely folded, carried nearly horizontal (see illustration, figure 39); sickle feathers, narrow, short, tapering; coverts, narrow, fine, short.

BREAST: Broad, rounded at sides.

BODY AND STERN: Body, fine and close on under part; not deep; stern, well tucked up underneath.

LEGS AND TOES: Thighs, long, muscular, standing out from body, but slightly sloping to hocks; shanks, long, smooth, bony, standing well apart; toes, long, straight, well spread.

PLUMAGE: Short, close, hard and firm.

STATION: Erect.

SIZE: Exceptionally large birds are undesirable.

Note: Cockerels shown after November first should be dubbed.

SHAPE OF FEMALE.

HEAD: Long, lean, bony.

FACE: Lean, thin with fine skin.

BEAK: Long, tapering, slightly curved.

EYES: Large, full, with keen expression.

COMB: Single, small, straight, thin, erect, evenly serrated.

WATTLES AND EAR-LOBES: Wattles, small, thin, round. Ear-lobes, small.

NECK: Long, very slightly arched, carried erect, tapering neatly and gradually from body to head, thin and clean-cut at throat, giving a distinct outline to head; neck feathers, lower portion, short, close.

WINGS: Large, powerful, the front standing out from body at shoulders, the feathers folded closely together, the points not extending beyond body; carried without drooping, but not carried over back.

BACK: Flat, rather short, straight on top from base of neck to tail, broad at shoulders, narrowing and sloping to stern.

TAIL: Rather short, compact, closely folded, carried nearly horizontal. (See illustration, figure 39.)

BREAST: Broad, rounded at sides.

BODY AND STERN: Body, fine and close on under part; not deep; stern, well tucked up underneath.

LEGS AND TOES: Thighs, long, muscular, standing out from body, but slightly sloping to hocks; shanks, long, smooth, bony, standing well apart; toes, long, straight, well spread.

PLUMAGE: Short, close, hard and firm.

STATION: Erect.

SIZE: Exceptionally large birds are undesirable.

BLACK-BREASTED RED GAMES.

COLOR OF MALE.

HEAD: Plumage, light orange.

BEAK: Horn.

EYES: Red.

COMB, FACE, WATTLES AND EAR-LOBES: Red.

NECK: Hackle, light golden; plumage in front of hackle, black.

WINGS: Shoulders, black; fronts, black; bows, red; coverts, lustrous black, forming a distinct bar across wing; primaries, black, except lower feather, the outer web of which should be bay; secondaries, part of outer webs forming wing-bay, bay; remainder of feathers, black.

BACK: Bright red; saddle, light golden.

TAIL: Black, sickle feathers and tail-coverts, lustrous black.

BREAST: Black.

BODY AND STERN: Black.

LEGS AND TOES: Thighs, black; shanks and toes, willow green.

COLOR OF FEMALE.

HEAD: Plumage, gold.

BEAK: Horn.

EYES: Red.

COMB, FACE, WATTLES AND EAR-LOBES: Red.

NECK: Light golden, with black stripe through middle of each feather, terminating in a point near its lower extremity; feathers in front of neck, reddish salmon.

WINGS: Shoulders, fronts, bows, coverts and secondaries, grayish-brown, stippled with golden-brown; primaries, black.

BACK: Grayish-brown, stippled with golden-brown.

TAIL: Black, except the two top feathers, which, with the coverts, should be brown.

BREAST: Light salmon, shading off to ashy-brown toward thighs.

BODY AND STERN: Ashy-brown.

LEGS AND TOES: Thighs, ashy-brown; shanks and toes, willow green.

BLACK-BREASTED RED GAME MALE

BLACK-BREASTED RED GAME FEMALE

BROWN-RED GAMES.

COLOR OF MALE.

HEAD: Plumage, orange.

BEAK: Black.

EYES: Black.

COMB, FACE, WATTLES AND EAR-LOBES: Dark purple.

NECK: Hackle, lemon, with a narrow, dark stripe through the middle of each feather, terminating in a point near its lower extremity; plumage in front of hackle, black laced with lemon.

WINGS: Shoulders, black; fronts, black; bows, lemon; coverts, lustrous black; primaries and secondaries, black.

BACK: Lemon; saddle, lemon, with a narrow dark stripe through the middle of each feather, terminating in a point near its lower extremity.

TAIL: Black; sickle feathers and tail coverts, lustrous black.

BREAST: Black, the feathers laced with lemon.

BODY AND STERN: Black.

LEGS AND TOES: Thighs, black; shanks and toes, black.

COLOR OF FEMALE.

HEAD: Plumage, lemon.

BEAK: Black.

EYES: Black.

COMB, FACE, WATTLES AND EAR-LOBES: Dark purple.

NECK: Lemon, with a narrow, dark stripe through middle of each feather, terminating in a point near its lower extremity; feathers in front of neck, black, each feather laced with lemon.

WING: Black.

BACK: Black.

TAIL: Black.

BREAST: Black, each feather laced with lemon.

BODY AND STERN: Black.

LEGS AND TOES: Thighs, black; shanks and toes, black.

GOLDEN DUCKWING GAMES.

COLOR OF MALE.

HEAD: Plumage, creamy white.

BEAK: Horn.

EYES: Red.

COMB, FACE, WATTLES AND EAR-LOBES: Red.

NECK: Hackle, creamy white, free from striping; plumage in front of hackle, black.

WINGS: Shoulders, black; fronts, black; bows, golden; greater and smaller coverts, blue-black, forming a distinct bar across wing; primaries, black, except lower feathers, outer web of which should be creamy-white; secondaries, part of outer webs forming wing-bay, creamy-white, remainder of feathers, black.

BACK: Golden; saddle, light golden, free from black striping.

TAIL: Black; sickle feathers and tail-coverts, lustrous blue-black; smaller coverts, light golden.

BREAST: Black.

BODY AND STERN: Black.

LEGS AND TOES: Thighs, black; shanks and toes, willow.

COLOR OF FEMALE.

HEAD: Plumage, silvery-gray.

BEAK: Horn.

EYES: Red.

COMB, FACE, WATTLES AND EAR-LOBES: Red.

NECK: Silvery-gray, with narrow, dark stripe through middle of each feather, terminating in a point near its lower extremity; feathers in front of neck, reddish-salmon.

WINGS: Shoulders, fronts, bows, coverts and secondaries, gray, stippled with dark gray; primaries, dark brown.

BACK: Gray, stippled with darker gray.

TAIL: Black, except the two top feathers, which should be gray, stippled with a darker gray.

BREAST: Rich salmon.

BODY AND STERN: Ashy-gray.

LEGS AND TOES: Thighs, ashy-gray; shanks and toes, willow.

SILVER DUCKWING GAME MALE

SILVER DUCKWING GAME FEMALE

SILVER DUCKWING GAMES.

COLOR OF MALE.

HEAD: Plumage, white.

BEAK: Horn.

EYES: Red.

COMB, FACE, WATTLES AND EAR-LOBES: Red.

NECK: Hackle, white, free from black stripes; plumage in front of hackle, black.

WINGS: Shoulders, black; fronts, black; bows, white; coverts, blue-black, forming a distinct bar across wings; primaries, black, except lower feathers, outer web of which should be white; secondaries, part of outer webs forming wing-bays, white, remainder of feathers, black.

BACK: White; saddle, white, free from black stripes.

TAIL: Black; sickle feathers and coverts, lustrous blue-black; smaller coverts, white.

BREAST: Black.

BODY AND STERN: Black.

LEGS AND TOES: Thighs, black; shanks and toes, willow.

COLOR OF FEMALE.

HEAD: Plumage, silvery-gray.

BEAK: Horn.

EYES: Red.

COMB, FACE, WATTLES AND EAR-LOBES: Red.

NECK: Silvery-gray, with a narrow, black stripe through middle of each feather, terminating in a point near its lower extremity; feathers in front of neck, dark salmon.

WINGS: Shoulders, fronts, bows, coverts and secondaries, light gray, finely stippled with darker gray; primaries, black.

BACK: Light gray, finely stippled with darker gray.

TAIL: Black, except the two top feathers, which should be light gray, stippled with darker gray.

BREAST: Light salmon.

BODY AND STERN: Ashy-gray.

LEGS AND TOES: Thighs, ashy-gray; shanks and toes, willow.

BIRCHEN GAMES.

COLOR OF MALE.

HEAD: Plumage, white.

BEAK: Black.

EYES: Black.

COMB, FACE, WATTLES AND EAR-LOBES: Dark purple.

NECK: Hackle, white, with narrow, dark stripe through middle of each feather, terminating in a point near its lower extremity; plumage in front of hackle, black, laced with white.

WINGS: Shoulders, black; fronts, black; bows, white; coverts, glossy black; primaries and secondaries, black.

BACK: White; saddle, white, with narrow, black stripe through middle of each feather.

TAIL: Black; sickle feathers and coverts, lustrous black.

BREAST: Ground color, black, the feathers laced with white.

BODY AND STERN: Black.

LEGS AND TOES: Thighs, black; shanks and toes, black.

COLOR OF FEMALE.

HEAD: Plumage, white.

BEAK: Black.

EYES: Black.

COMB, FACE, WATTLES AND EAR-LOBES: Dark purple.

NECK: White, with narrow, dark stripe through middle of each feather, terminating in a point near its extremity; feathers in front of neck, black, laced with white.

WINGS: Black.

BACK: Black.

TAIL: Black.

BREAST: Black, feathers laced with white.

BODY AND STERN: Black.

LEGS AND TOES: Thighs, black; shanks and toes, black.

271

RED PYLE GAME MALE

RED PYLE GAME FEMALE

RED PYLE GAMES.

COLOR OF MALE.

HEAD: Plumage, bright orange.

BEAK: Yellow.

EYES: Red.

COMB, FACE, WATTLES AND EAR-LOBES: Red.

NECK: Light orange; plumage in front of neck, white, which may be tinged with rich yellow.

WINGS: Shoulders, white; fronts, white; bows, red; coverts, white, forming a distinct bar across wings; primaries, white, except lower feathers, outer webs of which are bay; secondaries, part of outer webs forming the wing-bays, red, remainder of feathers, white.

BACK: Red; saddle, light orange.

TAIL: Main tail, sickle and coverts, white.

BREAST: White.

BODY AND STERN: White.

LEGS AND TOES: Thighs, white; shanks and toes, yellow.

COLOR OF FEMALE.

HEAD: Plumage, golden.

BEAK: Yellow.

EYES: Red.

COMB, FACE, WATTLES AND EAR-LOBES: Red.

NECK: White, the feathers edged with gold; feathers in front of neck white tinged with salmon.

WINGS: White.

BACK: White.

TAIL: White.

BREAST: Salmon.

BODY AND STERN: White.

LEGS AND TOES: Thighs, white; shanks and toes, yellow.

WHITE GAMES.

COLOR OF MALE AND FEMALE.

BEAK: Yellow.
EYES: Red.
COMB, FACE, WATTLES AND EAR-LOBES: Red.
LEGS AND TOES: Yellow.
PLUMAGE: Web, fluff and quills of feathers in all sections, pure white.

BLACK GAMES.

COLOR OF MALE AND FEMALE.

BEAK: Black.
EYES: Brown.
COMB, FACE, WATTLES AND EAR-LOBES: Deep red.
LEGS AND TOES: Black.
PLUMAGE: Surface, lustrous greenish-black throughout; under-color, dull black.

GAME BANTAMS.

Disqualifications.

Cocks not dubbed; artificial coloring; trimming or plucking of feathers; duck-foot. (See general disqualifications.)

STANDARD WEIGHTS.

Cock ....22 oz.		Hen ....20 oz.	
Cockerel ....20 oz.		Pullet ....18 oz.	

SHAPE AND COLOR OF MALE AND FEMALE.

The shape and color of Game Bantams shall be the same as the corresponding varieties of Games.

BLACK-BREASTED RED GAME BANTAM MALE

BLACK-BREASTED RED GAME BANTAM FEMALE

ORIENTALS.

Breeds	Varieties
SUMATRAS ....Black	
MALAYS ....Black-Breasted Red	
MALAY BANTAMS ....Black-Breasted Red	

SCALE OF POINTS FOR SUMATRAS.

Symmetry	4
Size	4
Condition	5
Comb—Shape 3, Color 3..............................	6
Head and Beak—Shape 2, Color 3.....................	5
Eyes—Shape 2, Color 2..............................	4
Ear-Lobes and Wattles—Shape 2, Color 2.............	4
Neck—Shape 4, Color 4..............................	8
Wings—Shape 4, Color 4.............................	8
Back—Shape 6, Color 4..............................	10
Tail—Shape 10, Color 4.............................	14
Breast—Shape 4, Color 4............................	8
Body and Stern—Shape 4, Color 4....................	8
Legs and Toes—Shape 4, Color 4.....................	8
Length of Feathers.................................	4
	———
	100

BLACK SUMATRAS.

The Black Sumatras are natives of Sumatra. This is a fowl with graceful form and lustrous greenish-black plumage throughout. Its peculiar characteristic is a long, drooping tail which has an abundance of smaller sickles and coverts. The Black Sumatra is very rare and comparatively few are bred in America.

Disqualifications.

White ear-lobes, feathers other than black in any part of plumage. (See general disqualifications.)

SHAPE OF MALE.

HEAD: Short and round.

BEAK: Of medium length, stout and well curved.

EYES: Large and bold.

COMB: Pea; small.

WATTLES AND EAR-LOBES: Very small.

NECK: Rather long, well arched; hackle, long, flowing.

WINGS: Long, large, carried with fronts slightly raised, points of feathers folded closely together, not drooping and not carried over the back.

BACK: Long, broad at shoulders, narrowing slightly to tail. with very long, flowing saddle feathers.

TAIL: Long, drooping, carried at an angle of thirty degrees above the horizontal (see illustration, figure 39), with abundance of feathers and coverts, which should be long and flowing.

BREAST: Broad, full.

BODY AND STERN: Body, firm, muscular, tapering to tail; stern compact.

LEGS AND TOES: Thighs, of medium length, large, strong; shanks, rather short, standing well apart; toes, long, straight, well spread.

SHAPE OF FEMALE.

HEAD: Short and round.

BEAK: Of medium length, stout and well curved.

EYES: Large and bold.

COMB: Pea; small.

WATTLES AND EAR-LOBES: Very small.

NECK: Rather long; feathers, long.

WINGS: Long, large, points not drooping and not carried over back.

BLACK: Long, broad at shoulders, narrowing slightly to tail.

TAIL: Long, large, drooping, carried at an angle of thirty degrees above the horizontal. (See illustration, figure 39.)

BREAST: Broad, round, full.

BODY AND STERN: Body, firm, muscular, tapering to tail; stern compact.

LEGS AND TOES: Thighs, of medium length, large, strong; shanks, rather short, standing well apart; toes, well spread.

COLOR OF MALE AND FEMALE.

BEAK: Black or dark slate.

EYES: Dark brown.

COMB AND FACE: Purple.

WATTLES AND EAR-LOBE: Dark red.

LEGS AND TOES: Black or dark slate.

PLUMAGE: Very lustrous greenish-black throughout.

UNDER-COLOR OF ALL SECTIONS: Dull black.

MALAYS.

The Malay is originally from India but was later perfected in Sumatra or Java. It is a fowl of moderate size, with long neck, rather long back slanting to tail which is carried below the horizontal. The body is firm and muscular. It is a tight feathered fowl.

SCALE OF POINTS.

Station 10
Weight and Height................................. 12
Condition 5
Head, Beak and Eyes—Shape 5, Color 6.............. 11
Comb 8
Wattles and Ear-Lobes—Shape 2, Color 2............ 4
Neck—Shape 3, Color 3............................. 6
Wings—Shape 4, Color 4............................ 8
Back—Shape 3, Color 3............................. 6
Tail—Shape 3, Color 3............................. 6
Breast—Shape 3, Color 3........................... 6
Body and Stern—Shape 2, Color 2................... 4
Thighs and Shanks—Shape 3, Color 3................ 6
Feet 2
Hardness of Feathers—Condition 3, Hardiness 3..... 6
                                                 ———
                                                 100

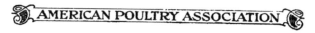

BLACK-BREASTED RED MALAYS.

STANDARD WEIGHTS.

Cock ....9 lbs.   Hen ....7 lbs.
Cockerel ....7 lbs.   Pullet ....5 lbs.

STANDARD HEIGHTS.

Cock ....26 in.   Hen ....18 in.
Cockerel ....18 in.   Pullet ....15 in.

Disqualifications.

Lopped combs; shanks or feet other than yellow; duck feet (See general disqualifications.)

SHAPE OF MALE.

HEAD: Broad, long, crown heavy and projecting over eyes, giving a fierce, cruel expression.

BEAK: Thick, short, strong.

EYES: Large.

COMB: Rather small, resembling a strawberry in front, set well forward. If males are dubbed, deduct eight points.

WATTLES AND EAR-LOBES: Very small.

NECK: Long, nearly straight and upright; hackle, short, scanty.

WINGS: Rather short, large, stout, bony, very prominent at shoulders, carried compactly against sides; wing-points resting under saddle feathers, without drooping or being carried over back.

BACK: Rather long, slanting, rather convex in outline, tapering to tail, large and broad at shoulders; saddle, narrow drooping; saddle feathers, short, scanty.

TAIL: Of medium length, drooping, carried below horizontal (see illustration, figure 39); well folded together; sickle feathers, curved, but not long.

BREAST: Broad, full; breast-bone, deep, prominent.

BODY AND STERN: Body, firm, muscular, broad at intersection of shoulders, tapering toward tail; stern, well tucked up.

THIGHS AND SHANKS: Thighs, long, hard, round, set well apart; shanks, long.

FEET: Flat, broad; toes, well apart, long, straight.

HARDNESS OF FEATHERS: Feathers, short, lying close, hard, firm and strong.

SHAPE OF FEMALE.

HEAD: Broad, long, crown heavy and projecting over eyes giving a fierce, cruel expression; face and throat, bare of feathers.

BEAK: Thick, short.

EYES: Large.

COMB: Rather small, resembling a strawberry in front, set well forward.

WATTLES AND EAR-LOBES: Very small.

NECK: Long, nearly straight and upright; feathers, short.

WINGS: Rather short, large, strong, bony, very prominent at shoulders, carried compactly against sides; wing-points resting under saddle feathers, without drooping or being carried over back.

BACK: Rather long, slanting, rather convex in outline, tapering to tail, large and broad at shoulders.

TAIL: Of medium length, drooping, carried slightly below the horizontal (see illustration, figure 39); closely carried.

BREAST: Broad, full; breast-bone, deep, prominent.

BODY AND STERN: Body, firm, muscular, broad at intersection of shoulders, tapering toward tail; stern, well tucked up.

THIGHS AND SHANKS: Thighs, long, hard, round, set well apart; shanks, long, bony, strong, standing evenly apart; scales, smooth.

FEET: Flat, broad; toes, well apart, long, straight.

HARDNESS OF FEATHERS: Feathers, short, lying close, hard, firm and strong.

COLOR OF MALE.

HEAD: Plumage, red or maroon.

BEAK: Yellow, or yellow striped with horn.

EYES: Pearl or yellow.

COMB, FACE, WATTLES AND EAR-LOBES: Rich red.

NECK: Hackle, dark red, shading into reddish-maroon; plumage other than hackle, black.

WINGS: Fronts, black; bows, a rich, glossy, dark red; coverts, glossy greenish-black, forming a wide bar across the wings; primaries, black, outer webs rich bay; secondaries, part of outer webs forming wing-bays, rich bay, the remainder of feathers black.

BACK: Rich, glossy, dark red or maroon; saddle feathers, rich, dark red.

TAIL: Black; sickle feathers and coverts, rich greenish-black.

BREAST: Glossy black.

283

BODY AND STERN: Black.

THIGHS AND SHANKS: Thighs, black; shanks and feet, yellow.

UNDER-COLOR OF ALL SECTIONS: Slate, tinged with brown.

COLOR OF FEMALE.

HEAD: Plumage, dark brown.

BEAK: Yellow, or yellow and horn.

EYES: Pearl or yellow.

COMB, FACE, WATTLES AND EAR-LOBES: Rich red.

NECK: Dark brown or brown striped with black; feathers in front of neck, cinnamon brown.

WINGS: Fronts, bows, coverts and secondaries, brown; primaries, very dark brown.

BACK: Dark or cinnamon brown.

TAIL: Very dark brown or black.

BREAST: Cinnamon brown.

BODY AND STERN: Brown.

THIGHS AND SHANKS: Thighs, brown; shanks and feet, yellow.

UNDER-COLOR OF ALL SECTIONS: Slate, tinged with brown.

BLACK-BREASTED RED MALAY BANTAMS.

STANDARD WEIGHTS.

Cock	26 oz.	Hen	24 oz.
Cockerel	24 oz.	Pullet	22 oz.

Disqualifications.

Same as for large Malays. (See general disqualifications.)

SHAPE AND COLOR OF MALE AND FEMALE.

The general shape and color of Black-Breasted Red Malay Bantams shall be the same as those of the Standard-size Malays.

ORNAMENTAL BANTAMS.

Breeds	*Varieties*
SEBRIGHTS	Golden / Silver
ROSE-COMB	White / Black
BOOTED	White
BRAHMAS	Light / Dark
COCHIN	Buff / Partridge / White / Black
JAPANESE	Black-Tailed / White / Black / Gray
POLISH	Bearded White / Buff-Laced / Non-Bearded
MILLE FLEUR	Booted

ORNAMENTAL BANTAMS.

Ornamental Bantams are bred chiefly for pleasure and fancy but they possess useful qualities as well. Among Ornamental Bantams, the Cochin and Brahma Bantams should be miniatures of the large Cochins and Brahmas in shape and color; the Black and White Rose-Combs should be counterparts of the graceful and stylish Hamburgs, carrying, however, wings and tails somewhat larger in proportion to the body. They are the embodiment of grace, style and sprightliness.

Sebright Bantams were originated in England and were the result of thirty years of painstaking care in mating and breeding. They are, perhaps, the greatest achievement of the fanciers' art in producing specimens of both sexes that are marvels of diminutive size and laced feathers in all sections. The feathers of both male and female are laced exactly alike. The low carriage of wings and well spread tail give to these beautiful specimens a most distinctive and striking appearance.

Japanese Bantams constitute one of the curiosities of the Bantam class. The disproportionately large comb, face, wings and tail of the male and remarkable shortness of legs are the chief characteristics. The tail is distinguished by the long sword-shaped sickles that are carried forward and upright to an unusual degree.

Polish Bantams should be the same shape and plumage as large Polish varieties.

The White Booted are distinct from the White Cochin Bantams in that they possess an abundance of stiff feathers, pronounced vulture-hocks, and an upright and sprightly carriage.

The Mille Fleur variety was introduced into this country in 1911 and is, therefore, comparatively new, but in England, Germany and Belgium is well known. It is a tri-colored variety of great beauty.

SEBRIGHT BANTAMS.

SCALE OF POINTS FOR SEBRIGHT AND ROSE-COMB BANTAMS.

Symmetry	4
Weight	2
Condition	4
Comb	8
Head—Shape 2, Color 2	4
Beak—Shape 2, Color 2	4
Eyes—Shape 2, Color 2	4
Wattles and Ear-Lobes—Shape 4, Color 4	8
Neck—Shape 4, Color 4	8
Wings—Shape 4, Color 6	10
Back—Shape 4, Color 6	10
Tail—Shape 6, Color 6	12
Breast—Shape 5, Color 5	10
Body and Fluff—Shape 4, Color 4	8
Legs and Toes—Shape 2, Color 2	4
	100

Disqualifications.

Cocks or cockerels having hackle feathers extending over shoulders, or sickle feathers extending more than an inch and a half beyond tail proper; shanks other than slaty-blue. (See general disqualifications.)

STANDARD WEIGHTS.

Cock	26 oz.	Hen	22 oz.
Cockerel	22 oz.	Pullet	20 oz.

SHAPE OF MALE.

HEAD: Large, round in front and carried well back.

BEAK: Short, slightly curved.

EYES: Large, round.

COMB: Rose; square in front, firm and even on head, terminating at rear in a spike, which inclines upward very slightly; top covered with small points; free from hollow center.

WATTLES AND EAR-LOBES: Wattles, broad, well rounded. Ear-lobes, smooth.

NECK: Tapering, well arched, carried very far back; hen-feathered; free from hackle feathers.

WINGS: Large, carried low, but not so low as to conceal hocks.

BACK: Very short, free from saddle hangers.

TAIL: Full, well expanded, carried at an angle of seventy degrees above the horizontal (see illustration, fig. 39); free from sickles; feathers broadest towards the ends, the two upper, which may be slightly curved, not extending more than an inch and a half beyond the others; coverts, straight, round at ends and lying close to sides of tail.

BREAST: Full, round, carried prominently forward.

BODY AND FLUFF: Body, compact, deep, short; fluff, short.

LEGS AND TOES: Thighs, very short, stout; shanks, short, rather slender; toes, straight.

SHAPE OF FEMALE.

HEAD: Broad and well rounded.

BEAK: Short, slightly curved.

EYES: Large, round.

COMB: Rose; similar to that of male, but very small.

WATTLES AND EAR-LOBES: Wattles, small, well rounded. Ear-lobes, flat, smooth, small.

NECK: Tapering, very slightly arched.

WINGS: Large, carried low, but not so low as to conceal hocks.

BACK: Short, tapering to tail.

TAIL: Full, well expanded, carried at an angle of seventy degrees above the horizontal. (See illustration, fig. 39.)

BREAST: Full, round, carried prominently forward.

BODY AND FLUFF: Body, compact, deep, short; fluff, short.

LEGS AND TOES: Thighs, very short, stout; shanks, short, rather slender; toes, straight.

288

GOLDEN SEBRIGHT BANTAMS.

COLOR OF MALE.

BEAK: Dark horn.

EYES: Brown.

COMB AND FACE: Reddish-purple.

WATTLES AND EAR-LOBES: Wattles, bright red. Ear-lobes, reddish-purple preferred.

SHANKS AND TOES: Slaty-blue.

PLUMAGE: Surface throughout, golden-bay, each feather evenly and distinctly laced all around with a narrow edging of lustrous black.

UNDER-COLOR OF ALL SECTIONS: Slate.

COLOR OF FEMALE.

BEAK: Dark horn.

EYES: Brown.

COMB AND FACE: Reddish-purple.

WATTLES AND EAR-LOBES: Wattles, bright red. Ear-lobes, reddish-purple preferred.

SHANKS AND TOES: Slaty-blue.

PLUMAGE: Surface throughout, golden-bay, each feather evenly and distinctly laced all around with a narrow edging of lustrous black.

UNDER-COLOR OF ALL SECTIONS: Slate.

SILVER SEBRIGHT BANTAM MALE

SILVER SEBRIGHT BANTAM FEMALE

SILVER SEBRIGHT BANTAMS.

COLOR OF MALE.

BEAK: Dark horn.

EYES: Brown.

COMB AND FACE: Reddish-purple.

WATTLES AND EAR-LOBES: Wattles, bright red. Ear-lobes, reddish-purple preferred.

SHANKS AND TOES: Slaty-blue.

PLUMAGE: Surface throughout, silvery-white, each feather evenly and distinctly laced all around with a narrow edging of lustrous black.

UNDER-COLOR OF ALL SECTIONS: Slate.

COLOR OF FEMALE.

BEAK: Dark horn.

EYES: Brown.

COMB AND FACE: Reddish-purple.

WATTLES AND EAR-LOBES: Wattles, bright red. Ear-lobes, reddish-purple preferred.

SHANKS AND TOES: Slaty-blue.

PLUMAGE: Surface throughout, silvery-white, each feather evenly and distinctly laced all around with a narrow edging of lustrous black.

UNDER-COLOR OF ALL SECTIONS: Slate.

ROSE-COMB BANTAMS.

STANDARD WEIGHTS.

Cock ....26 oz. Hen ....22 oz.
Cockerel ....22 oz. Pullet ....20 oz.

SHAPE OF MALE.

HEAD: Small, round, carried well backward over the body.

BEAK: Short, slightly curved.

EYES: Full.

COMB: Rose; square in front, firm and even on head, terminating at rear in spike, which inclines upward very slightly; top covered with small points; free from hollow center.

WATTLES AND EAR-LOBES: Wattles, broad, thin, well rounded. Ear-lobes, prominent, flat, round, smooth, even, fitted closely to head.

NECK: Tapering, carried back so as to bring head toward tail; hackle, full, long, descending well over shoulders.

WINGS: Large, points carried low but not to conceal hocks; secondaries, slightly expanded.

BACK: Very short, tapering to tail; saddle feathers, abundant, long.

TAIL: Full, well expanded, carried at an angle of forty degrees above the horizontal (see illustration, figure 39); sickles, long, well curved; coverts, abundant.

BREAST: Full, round, carried prominently forward.

BODY AND FLUFF: Body, plump, tapering toward tail; fluff, short.

LEGS AND TOES: Thighs, short; shanks, smooth, short, tapering; toes straight.

SHAPE OF FEMALE.

HEAD: Small, round.

BEAK: Short, slightly curved.

EYES: Full.

COMB: Rose; similar to that of male, but smaller.

WATTLES AND EAR-LOBES: Wattles, small, thin, well rounded. Ear-lobes, prominent, flat, round, smooth, even, fitting closely to head.

ROSE-COMB BLACK BANTAM MALE

ROSE-COMB BLACK BANTAM FEMALE

NECK: Short, tapering, carried well back.

WINGS: Large, points carried low, but not to conceal hocks.

BACK: Short, tapering to tail.

TAIL: Full, well expanded, carried at an angle of forty degrees above the horizontal. (See illustration, figure 39.)

BREAST: Full, round, carried prominently forward.

BODY AND FLUFF: Body, compact, tapering toward tail; fluff, short.

LEGS AND TOES: Thighs, short; shanks, smooth, short, tapering; toes, straight.

ROSE-COMB WHITE BANTAMS.

Disqualifications.

Feathers other than white in any part of plumage; shanks other than white. (See general disqualifications.)

COLOR OF MALE AND FEMALE.

BEAK: White.

EYES: Reddish-bay.

COMB, FACE AND WATTLES: Bright red.

EAR LOBES: White.

SHANKS AND TOES: White, with pinkish tinge on back of shanks and between scales.

PLUMAGE: Web, fluff and quills of feathers in all sections, pure white.

ROSE-COMB BLACK BANTAMS.

Disqualifications.

Pure white in any part of plumage, extending over half an inch, or two or more feathers tipped or edged with positive white; shanks other than black or very dark leaden-blue. (See general disqualifications.)

COLOR OF MALE AND FEMALE.

BEAK: Black.

EYES: Brown.

COMB, FACE AND WATTLES: Bright red. Ear-lobes, white.

SHANKS AND TOES: Black.

PLUMAGE: Surface throughout, lustrous greenish-black.

UNDER-COLOR OF ALL SECTIONS: Dull black,

SCALE OF POINTS FOR BOOTED WHITE, BRAHMA, COCHIN, JAPANESE AND MILLE FLEUR BANTAMS.

Symmetry	4
Weight	2
Condition	4
Comb	8
Head—Shape 2, Color 2	4
Beak—Shape 2, Color 2	4
Eyes—Shape 2, Color 2	4
Wattles and Ear-Lobes—Shape 2, Color 2	4
Neck—Shape 4, Color 6	10
Wings—Shape 4, Color 6	10
Back—Shape 6, Color 4	10
Tail—Shape 5, Color 5	10
Breast—Shape 5, Color 5	10
Body and Fluff—Shape 5, Color 3	8
Legs and Toes—Shape 5, Color 3	8
	100

BOOTED WHITE BANTAMS.

Disqualifications.

Feathers other than white in any part of plumage; absence of vulture hocks; shanks not feathered down the outer sides; shanks other than white; outer toes not feathered. (See general disqualifications.)

STANDARD WEIGHTS.

Cock	26 oz.	Hen	22 oz.
Cockerel	22 oz.	Pullet	20 oz.

SHAPE OF MALE.

HEAD: Small, round, carried well back.

BEAK: Short, slightly curved.

EYES: Full.

COMB: Single; of medium size, firm and straight on head, with five well defined points, evenly serrated.

WATTLES AND EAR-LOBES: Wattles, broad, thin, well rounded. Ear-lobes, smooth.

NECK: Tapering, curved well back, with full, long hackle, descending well over shoulders.

WINGS: Large, points carried a little low.

BACK: Short; saddle feathers, abundant, long.

TAIL: Full, well expanded; carried at an angle of ninety degrees above the horizontal (see illustration, figure 39); sickles, long, well curved; coverts, abundant, long.

BREAST: Full, round.

BODY AND FLUFF: Body, rather short and compact; fluff, moderately full.

LEGS AND TOES: Thighs, long, well furnished with long, stiff feathers or vulture-hocks, which almost touch the ground; shanks, long, heavily feathered on the outer sides; toes, straight; outer toes, heavily feathered on their extremities.

SHAPE OF FEMALE.

HEAD: Small, round.

BEAK: Short, slightly curved.

EYES: Full

COMB: Single, small, firm and straight on head, with five well defined points, evenly serrated.

WATTLES AND EAR-LOBES: Wattles, small, well rounded. Ear-lobes, flat.

NECK: Of medium length, tapering, carried well back.

WINGS: Large, points carried a little low.

BACK: Short.

TAIL: Full, well expanded, carried at an angle of seventy-five degrees above the horizontal. (See illustration, figure 39.)

BREAST: Full, round.

BODY AND FLUFF: Body, rather short and compact; fluff, moderately full.

LEGS AND TOES: Thighs, long, well furnished with long, stiff feathers or vulture hocks; shanks, long, heavily feathered on outer sides; toes, straight; outer toes, heavily feathered to their extremities.

COLOR OF MALE AND FEMALE.

BEAK: White.

EYES: Reddish-bay.

COMB, FACE, WATTLES AND EAR-LOBES: Bright red.

SHANKS AND TOES: White.

PLUMAGE: Web, fluff and quill of feathers in all sections, pure white.

BRAHMA BANTAMS.

(Light and Dark.)

Brahma Bantams, male and female, should conform in miniature fashion to the general outlines of the larger Brahmas. Stiff hock plumage is very objectionable.

Disqualifications.

Disqualifications for Brahma Bantams shall be the same as for larger Brahmas of the corresponding variety, except in hock plumage and weight. (See general disqualifications.)

STANDARD WEIGHTS.

Cock ....30 oz.   Hen ....26 oz.
Cockerel ....26 oz.   Pullet ....24 oz.

SHAPE AND COLOR OF MALE AND FEMALE.

The general shape and color of the Brahma Bantams shall conform to the description of the corresponding variety of the larger Brahmas.

BUFF COCHIN BANTAM MALE

BUFF COCHIN BANTAM FEMALE

BLACK COCHIN BANTAM MALE

BLACK COCHIN BANTAM FEMALE

COCHIN BANTAMS.

The Cochin Bantam male should conform in miniature to the general outlines of the larger Cochin. He should be broad, deep, plump and well rounded, of bold and forward carriage, short in legs, head carried much higher than tail. Plumage, long, loose and abundant, the more fluff plumage the better. The Cochin Bantam female should conform, in a feminine way, with the male. In general outlines she should be rather short, neat and well rounded, very profuse of feathering and short in leg. Stiff hock plumage is objectionable in both male and female.

Disqualifications.

The disqualifications for all Cochin Bantams shall be the same as for larger Cochins of the corresponding varieties, except hock plumage and weight. (See general disqualifications.)

STANDARD WEIGHTS.

Cock ....30 oz.   Hen ....26 oz.
Cockerel ....26 oz.   Pullet ....24 oz.

SHAPE AND COLOR OF MALE AND FEMALE.

The general shape and color of all Cochin Bantams shall conform to the description of the corresponding variety of the larger Cochins.

JAPANESE BANTAMS.

STANDARD WEIGHTS.

Cock ....26 oz.   Hen ....22 oz.
Cockerel ....22 oz.   Pullet ....20 oz.

SHAPE OF MALE.

HEAD: Rather large and broad.

BEAK: Strong, well curved.

EYES: Large.

COMB: Single; large, firm and straight on head; evenly serrated, having five distinct points.

WATTLES AND EAR-LOBES: Wattles, large, pendant. Ear-lobes, large, smooth.

NECK: Rather short, curving prominently backward, with abundant hackle flowing well over shoulders.

WINGS: Large, long, points decidedly drooping.

BACK: Very short; saddle feathers, abundant.

TAIL: Very large, somewhat expanded, carried forward of perpendicular (see illustration, fig. 39), so as to almost come in contact with back of head; sickles, long, very upright, slightly curved.

BREAST: Very full, round, carried prominently forward.

BODY AND FLUFF: Body, rather short, deep and compact; fluff, short.

LEGS AND TOES: Thighs, of medium size, short; shanks, very short, smooth; toes, straight.

SHAPE OF FEMALE.

HEAD: Rather large and broad.

BEAK: Strong, well curved.

EYES: Large.

COMB: Single; large, firm and straight on head; evenly serrated, having five distinct points.

WATTLES AND EAR-LOBES: Wattles, of medium size, well rounded. Ear-lobes, of medium size, smooth.

NECK: Short, well curved.

WINGS: Large, long, points decidedly drooping.

BLACK-TAILED JAPANESE BANTAM MALE

BLACK-TAILED JAPANESE BANTAM FEMALE

BACK: Short.

TAIL: Large, somewhat expanded, carried forward of perpendicular. (See illustration, fig. 39.)

BREAST: Full, round, prominent.

BODY AND FLUFF: Body, rather short, deep and compact; fluff, short.

LEGS AND TOES: Thighs, of medium size, short; shanks, very short, smooth; toes, straight.

BLACK-TAILED JAPANESE BANTAMS.

Disqualifications.

Shanks other than yellow. (See general disqualifications.)

COLOR OF MALE.

HEAD: Plumage, white.

BEAK: Yellow.

EYES: Reddish-bay.

COMB, FACE, WATTLES AND EAR-LOBES: Bright red.

NECK: White; plumage in front of neck, white.

WINGS: Bows and coverts, white; primaries, black, edged with white; secondaries, black, with wide edging of white on upper webs, lower webs white; wing when folded shows white only.

BACK: White; saddle feathers, white.

TAIL: Black; sickles and coverts, black, edged with white.

BREAST: White.

BODY AND FLUFF: White.

LEGS AND TOES: Thighs, white; shanks and toes, yellow.

UNDER-COLOR OF ALL SECTIONS: White or bluish-white.

COLOR OF FEMALE.

HEAD: Plumage, white.

BEAK: Yellow.

EYES: Reddish-bay.

COMB, FACE, WATTLES AND EAR-LOBES: Bright red.

NECK: White; feathers in front of neck, white.

WINGS: Bows and coverts, white; primaries, black, edged with white; secondaries, black with wide edging of white on upper web, lower web white; wing, when folded, shows white only.

BACK: White.

TAIL: Black, two top feathers edged with white; coverts, black ged with white.
BREAST: White.
BODY AND FLUFF: White.
LEGS AND TOES: Thighs, white; shanks and toes, yellow.
UNDER-COLOR OF ALL SECTIONS: White or bluish-white.

WHITE JAPANESE BANTAMS.

Disqualifications.

Feathers other than white in any part of plumage; shanks other than yellow. (See general disqualifications.)

COLOR OF MALE AND FEMALE.

BEAK: Yellow.
EYES: Reddish-bay.
COMB, FACE, WATTLES AND EAR-LOBES: Bright red.
SHANKS AND TOES: Yellow.
PLUMAGE: Web, fluff and quill of feathers in all sections, pure white.

BLACK JAPANESE BANTAMS.

Disqualifications.

Pure white in any part of plumage extending over half an inch, or two or more feathers tipped or edged with positive white; shanks other than yellow or yellowish black. (See general disqualifications.)

COLOR OF MALE AND FEMALE

BEAK: Yellow, or yellow shaded with black.
EYES: Brown.
COMB, FACE, WATTLES AND EAR-LOBES: Bright red.
SHANKS AND TOES: Yellow, or yellow shaded with black.
PLUMAGE: Surface, lustrous greenish-black throughout.
UNDER-COLOR OF ALL SECTIONS: Dull black.

GRAY JAPANESE BANTAMS.

Disqualifications.

Shanks other than yellow or yellowish black. (See general disqualifications.)

COLOR OF MALE.

HEAD: Plumage, silvery-gray.

BEAK: Yellow shaded with dark horn.

EYES: Brown.

COMB, FACE, WATTLES AND EAR-LOBES: Bright red.

NECK: Hackle, silvery-gray with narrow, dark stripe terminating in a point near lower extremity of feather; plumage in front of neck, same as breast.

WINGS: Shoulders, black; fronts, black; bows, silvery white; coverts, glossy black; primaries and secondaries, black.

BACK: Silvery white; saddle, silvery white with narrow, dark stripe through middle of each feather, terminating in a point near its extremity.

TAIL: Black; sickles and tail coverts, glossy greenish-black.

BREAST: Ground color, black, feathers laced with silvery gray.

BODY AND FLUFF: Black.

LEGS AND TOES: Thighs, black; shanks and toes, yellow, or yellow shaded with black.

UNDER-COLOR OF ALL SECTIONS: Dark slate.

COLOR OF FEMALE.

HEAD: Plumage, white.

BEAK: Yellow, shaded with dark horn.

EYES: Brown.

COMB, FACE, WATTLES AND EAR-LOBES: Bright red.

NECK: White with narrow dark stripe through middle of each feather, terminating in a point near its lower extremity; feathers in front of neck, same as breast.

WINGS: Black.

BACK: Black.

TAIL: Black.

BREAST: Black, feathers laced with white.

BODY AND FLUFF: Black.

LEGS AND TOES: Thighs, black; shanks and toes, yellow, or yellow shaded with horn.

UNDER COLOR OF ALL SECTIONS: Dark slate.

POLISH BANTAMS.

SCALE OF POINTS FOR POLISH BANTAMS.

Symmetry 4
Weight 2
Condition 4
Comb 2
Head, Beak and Eyes — Shape 2, Color 2, each............. 12
Crest — Shape 10, Color 5.................................. 15
Wattles and Ear-Lobes 4, Beard 4......................... 8
Neck — Shape 3, Color 3................................... 6
Wings — Shape 4, Color 6.................................. 10
Back — Shape 4, Color 4................................... 8
Tail — Shape 4, Color 5................................... 9
Breast — Shape 4, Color 4................................. 8
Body and Fluff — Shape 3, Color 3........................ 6
Legs and Toes — Shape 3, Color 3......................... 6

————

100

STANDARD WEIGHTS.

Cock ....26 oz.   Hen ....22 oz.
Cockerel ....22 oz.   Pullet ....20 oz

Disqualifications.

See general disqualifications.

SHAPE AND COLOR OF MALE AND FEMALE.

The general shape and color of all Polish Bantams shall conform to the description of the corresponding variety of the larger Polish.

NON-BEARDED POLISH BANTAMS.

The same as the bearded varieties in every respect, including disqualifications, shape and color, except that they have no beard.

MILLE FLEUR BOOTED BANTAMS

Disqualifications.

Absence of vulture-hocks; shanks not feathered down the outer sides; outer toes not feathered. (See general disqualifications.)

STANDARD WEIGHT.

Cock ....26 oz.   Hen ....22 oz.
Cockerel ....22 oz.   Pullet ....20 oz.

SHAPE OF MALE.

HEAD: Small, round, carried well back.

BEAK: Short, slightly curved.

EYES: Full, round.

COMB: Single; of medium size, firm and straight on head. evenly serrated.

WATTLES AND EAR-LOBES: Wattles, broad, thin, well rounded. Ear-lobes, flat, almond shape.

BEARD: (In bearded varieties) thick and full, extending back of eye, projecting from sides of face.

NECK: Tapering, well arched, with full, long hackle, extending well over shoulders.

WINGS: Large, points carried rather low.

BACK: Short, forming a slight concave sweep to juncture of tail; saddle, long, abundant.

TAIL: Long, well spread, carried very erect; sickles, medium in length, slightly curved; coverts, abundant.

BODY AND FLUFF: Body, rather short and compact; fluff, moderately full.

LEGS AND TOES: Thighs, medium in length, covered with long, stiff feathers, forming a vulture hock almost touching the ground, shanks, medium in length, heavily feathered on outer sides; toes, straight; outer toes, heavily feathered to their extremities.

SHAPE OF FEMALE.

HEAD: Small, round, carried well back.

BEAK: Short, slightly curved.

EYES: Full, round.

COMB: Single; firm, small and straight on head, evenly serrated.

WATTLES AND EAR-LOBES: Wattles, small, well rounded. Ear-lobes, flat, almond shape.

BEARD: (In bearded varieties) thick and full, extending back of eye, projecting from sides of face.

NECK: Medium in length, well arched.

WINGS: Large, points carried rather low.

BACK: Short, forming a slight concave sweep to juncture of tail.

TAIL: Long, well spread, carried very erect.

BREAST: Full, round, carried well forward.

BODY AND FLUFF: Body, rather short and compact; fluff, moderately full.

LEGS AND TOES: Thighs, medium in length, covered with long, stiff feathers, forming a vulture-hock, almost touching the ground; shanks, medium in length, heavily feathered on outer sides; toes, straight; outer toes, heavily feathered to their extremities.

COLOR OF MALE.

HEAD: Plumage, bright red, each feather tipped with a small white spangle, a narrow bar of black dividing the white spangle from balance of feather.

BEAK: Horn.

EYES: Reddish-bay.

COMB, FACE, WATTLES AND EAR-LOBES: Bright red.

NECK: Plumage, rich, bright red, each feather having a greenish-black stripe extending lengthwise through lower part of feather, terminating near end of feather; feathers tipped with V-shaped spangle of pure white; plumage in front of neck, same color as breast; in bearded varieties, each feather of beard tipped with white, balance of feather, red.

WINGS: Bows, rich, brilliant red, each feather tipped with a pure white spangle; coverts, rich bay, each feather having a V-shaped white spangle at end of feather, a crescent-shaped bar of black dividing white spangle from balance of feather, which is rich bay, primaries, inner webs black, outer webs, black, slightly edged with bay, lower portion of feather edged with white; secondaries, outer webs, bay, inner webs, dull black, extending into outer webs near end of feathers, end of feathers tipped with white.

BACK: Rich bright red, each feather having a greenish-black stripe extending lengthwise through lower part of feather and terminating near end of feather, end of feather tipped with a V-shaped white spangle, the black stripe broadening out to edge of feather where it meets white spangle.

TAIL: Black, each feather tipped with white; coverts and sickles, greenish-black, each feather tipped with a V-shaped white spangle.

BREAST: Golden-bay, each feather tipped with a V-shaped white spangle, a crescent-shaped bar of black dividing white spangle from balance of feather.

BODY AND FLUFF: Body, plumage, same in color and markings as breast; fluff, dull black, mottled with white.

LEGS AND TOES: Thighs, plumage, same in colors and markings as breast; fluff, dull black, mottled with white; shanks and toes, slaty-blue; bottom of feet, yellow; shank and toe feathers, black, tipped with white.

UNDER-COLOR OF ALL SECTIONS: Slate, shading to grayish buff at base.

COLOR OF FEMALE.

HEAD: Plumage, rich, golden buff, each feather marked with crescent-shaped spangle of black near end of feather, end of feather tipped with a pure white V-shaped spangle.

BEAK: Horn.

EYES: Reddish-bay.

COMB, FACE, WATTLES AND EAR-LOBES: Bright red.

NECK: Plumage, rich, golden buff, each feather marked with a crescent-shaped spangle of black near end of feather, end of feather tipped with a pure white V-shaped spangle; in bearded varieties, each feather of beard tipped with white, balance of feather light golden buff; feathers in front of neck, same as breast.

WINGS: Bows and coverts, rich, golden buff, each feather marked with crescent-shaped spangle of black near end of feather, end of feather tipped with a pure white V-shaped spangle; primaries, inner webs, black, outer webs, black; secondaries, outer webs light golden buff extending nearly to end of feathers; inner webs dull black, extending nearly to end of feathers and broadening out into outer webs nearly to end of feathers where it joins a spangle of white at tips.

314

BACK: Rich, golden buff, each feather marked with a crescent-shaped spangle of black near end of feather, end of feather tipped with a pure white V-shaped spangle.

TAIL: Dull black, tipped with white; coverts, same as back.

BREAST: Same as back

BODY AND FLUFF: Same as back.

LEGS AND TOES: Thighs, same as back; hock, same as back; shanks and toes, slaty-blue; bottom of feet, yellow; shank and toe feathering, dull black, tipped with white.

UNDER-COLOR OF ALL SECTIONS: Slate, shading to grayish buff at base.

MISCELLANEOUS.

Breeds .... { Silkies
                                               Sultans
                                               Frizzles

SCALE OF POINTS FOR SILKIES.

Symmetry 4
Size 4
Condition 4
Head and Beak—Shape 3, Color 3........................... 6
Eyes—Shape 1, Color 2..................................... 3
Comb and Crest—Comb 5, Crest 10 15
Wattles and Ear-Lobes—Shape 2, Color 4.................. 6
Neck—Shape 4, Color 3..................................... 7
Wings—Shape 4, Color 6.................................... 10
Back—Shape 4, Color 3..................................... 7
Tail—Shape 4, Color 3..................................... 7
Breast—Shape 3, Color 3................................... 6
Body and Fluff—Shape 3, Color 3........................... 6
Legs and Toes—Shape 3, Color 3............................ 6
Texture of Plumage 9
                                                        ———
                                                        100

SILKIES.

Silkies derive their name from the peculiar formation of the plumage, their feathers being webless and of a silky texture. This peculiarity lends attractiveness to the breed, as it is possessed by no other Standard fowl. In general appearance, they are short, have silky feathered legs, broad backs and profuse plumage. The length of the webless feathers is a particularly desirable feature. The contrast formed by the white plumage with the purple face is noticeable on account of its singularity; in fact, the appearance of these specimens presents, throughout, a series of contradictions to the generally accepted laws which govern Standard-bred fowls.

Disqualifications.

Absence of crest or of fifth toe; feathers not truly silky; shanks not feathered down outer sides; vulture-hocks. (See general disqualifications.)

SHAPE OF MALE.

HEAD: Short, round.

BEAK: Short, stout.

EYES: Large.

COMB AND CREST: Comb, nearly round, crown covered with small corrugations, set prominently and firmly on the head; crest, soft and full, as upright as comb will permit, falling gracefully backward.

WATTLES AND EAR-LOBES: Wattles, of medium length; convex outer surface, nearly semi-circular. Ear-lobes, oval.

NECK: Short, with very full hackle, flowing well over shoulders.

WINGS: Small, carried low, the primaries and secondaries have a ragged fringe-like appearance and the ends are fairly covered by the saddle hangers.

BACK: Broad, short; rising gradually from about the middle of back toward tail.

TAIL: Small, main feathers having a ragged, fringe-like appearance.

BREAST: Full.

BODY AND FLUFF: Body, broad, squarely formed; fluff, full and abundant.

LEGS AND TOES: Thighs, short, well furnished with silky fluff; shanks, short, feathered on outer sides with silky plumage; toes, five on each foot, the outer toes feathered.

SIZE: Cock should not weigh more than three pounds.

SHAPE OF FEMALE.

HEAD: Small, short, round.

BEAK: Short, stout.

EYES: Large.

COMB AND CREST: Comb, similar to that of male, but very small. Crest, small, globular, erect.

WATTLES AND EAR-LOBES: Wattles, small. Ear-lobes, small and round.

NECK: Short, with abundant plumage.

WINGS: Small, carried low; the primaries and secondaries have a ragged, fringe-like appearance and the ends are fairly covered by the back plumage.

BACK: Broad and short, rising in a concave sweep from middle of back to a rounded cushion which extends to tail.

TAIL: Small, almost concealed by cushion and fluff; main tail feathers have a ragged, fringe-like appearance.

BREAST: Full.

BODY AND FLUFF: Body, broad, compact; fluff, full and abundant.

LEGS AND TOES: Thighs, short, well furnished with silky fluff; shanks, short, feathered on outer sides with silky plumage; toes, five on each foot, the outer toes, feathered.

PLUMAGE: Soft, silky, webless.

SIZE: Hens should not weigh more than two pounds.

COLOR OF MALE AND FEMALE.

BEAK: Blue.

EYES: Black.

COMB AND FACE: Purple.

WATTLES AND EAR-LOBES: Wattles, purple. Ear-lobes, light blue.

SHANKS AND TOES: Leaden-blue.

PLUMAGE: White.

SULTANS.

They have for their most attractive characteristics the novel features of a full crest, muff and beard, combined with vulture-hocks and profuse shank and toe feathering. These peculiarities should be most prominent in their form and outlines.

Disqualifications.

Beak other than white or pale flesh color; large, red face; absence of beard; absence of vulture-hocks; shanks not feathered down outer sides. (See general disqualifications.)

SCALE OF POINTS.

Symmetry	4
Size	4
Condition	5
Head — Shape 3, Color 3	6
Comb	4
Crest — Shape 8, Color 4	12
Wattles and Ear-Lobes 3, Beard 8, Whiskers 2	13
Neck — Shape 4, Color 3	7
Wings — Shape 3, Color 3	6
Back — Shape 4, Color 3	7
Tail — Shape 5, Color 3	8
Breast — Shape 3, Color 3	6
Body and Fluff — Shape 3, Color 3	6
Shanks and Toes — Shape 9, Color 3	12
	100

SHAPE OF MALE.

HEAD: Medium size.

BEAK: Short, well curved.

NOSTRILS: Large.

EYES: Oval.

COMB AND CREST: Comb, very small, having two spikes, V-shaped. Crest, large, globular, and compact.

BEARD: Very full, joining whiskers and extending to crest.

WATTLES AND EAR-LOBES: Wattles, small, round. Ear-lobes, small, round, concealed by crest and whiskers.

NECK: Short, arched, carried well back.

WINGS: Rather large, carried low.

BACK: Rather long, straight.

TAIL: Large, full, abundantly furnished with sickles and coverts.

BREAST: Deep and prominent.

BODY AND FLUFF: Body, very square, deep, compact, carried low.

LEGS AND TOES: Thighs, very short, well feathered, with long, full, vulture-hocks; shanks, short, heavily feathered down outer sides; toes, five on each foot, straight, the middle and outer toes well feathered.

SIZE: Cocks should weigh five pounds.

SHAPE OF FEMALE.

HEAD: Medium size.

BEAK: Short, well curved.

NOSTRILS: Large.

EYES: Oval.

COMB AND CREST: Comb, very small, having two spikes, V-shaped. Crest, large, globular and compact.

BEARD: Very full, joining whiskers and extending to crest.

WATTLES AND EAR-LOBES: Wattles, small, round. Ear-lobes, small, round, concealed by crest and whiskers.

NECK: Short, arched, carried well back.

WINGS: Large, carried low.

BACK: Long and straight.

TAIL: Large, well expanded, rather erect.

BREAST: Deep and prominent.

BODY AND FLUFF: Body, very square, compact, carried low.

LEGS AND TOES: Thighs, very short, well feathered with long vulture hocks; shanks, short, heavily feathered down outer sides; toes, five on each foot; straight middle and outer toes well feathered.

COLOR OF MALE AND FEMALE.

BEAK: Slaty-blue.

EYES: Reddish-bay.

COMB AND FACE: Comb, bright red. Face, bright red, but covered by whiskers and almost invisible.

WATTLES AND EAR-LOBES: Bright red.

SHANKS AND TOES: Slaty-blue.

PLUMAGE: Web, quill and fluff of feathers in all sections, pure white.

FRIZZLES.

Combs other than single; not matching in combs, in color of legs or in color of plumage when shown in pairs, trios or pens; more than four toes. (See general disqualifications.)

NO SCALE OF POINTS.

MALE AND FEMALE.

The feathers show a tendency to curve backward or upward at the ends, this curving at the ends being most noticeable in the hackle and saddle feathers, but the more all of the feathers are curved, the better. Feathers curving upward on neck and back of head, after the style of the hood in hooded pigeons, to be encouraged.

COLOR: Solid; black, white, red and bay admissible, provided the birds match when shown in pairs, trios and pens.

COMBS: Single.

CLASS XIII.

DUCKS.

Breeds	*Varieties*
PEKIN	White
AYLESBURY	White
ROUEN	Colored
CAYUGA	Black
CALL	Gray / White
EAST INDIA	Black
MUSCOVY	Colored / White
SWEDISH	Blue
BUFF	Buff
CRESTED	White
RUNNER	Fawn and White / White / Penciled

SCALE OF POINTS FOR ALL DUCKS EXCEPT RUNNER AND CRESTED WHITE.

Symmetry	4
Weight	6
Condition	10
Head — Shape 2, Color 2	4
Bill — Shape 2, Color 2	4
Eyes — Shape 2, Color 2	4
Neck — Shape 4, Color 4	8
Wings — Shape 4, Color 6	10
Back — Shape 8, Color 4	12
Tail — Shape 2, Color 2	4
Breast — Shape 10, Color 4	14
Body — Shape 12, Color 4	16
Legs and Toes — Shape 2, Color 2	4
	100

Note: For "weight," read, "smallness of size" in applying the scale to Call and East India Ducks.

DUCKS.

The male and female Pekin, Aylesbury, Rouen, Buff and Cayuga Ducks should be large, long and broad in body, full in breast, with deep, well set keels and all sections finely rounded, giving them a finished, plump appearance. The nearer they are bred to Standard weights the better. The head should be large, the bill broad and long on upper line, and the eyes bright. The neck should be of good length and well arched.

The body of White and Colored Muscovys should be long, broad and deep; the tail of good length, carried nearly horizontal; wings, large, long and powerful; shanks and feet short and large; head of good size, the top covered with crest-like feathers, which are elevated under excitement. The sides of face should be covered with caruncles—the larger, the better. The male in both varieties is fully one-third larger than the female.

The Runner Ducks are of moderate size and have a racy appearance. The neck should be long and thin; back, long, straight and narrow; breast, full, but not rounded, showing but little suggestion of keel; carriage, erect.

East India and Call Ducks—the bantams of the duck family—should be small in size, both male and female—the smaller, the better. The body should be short, well rounded and carried nearly horizontal. The head is small, the bill short and concave and the neck short and slender.

The crested Whites have the general characteristics of the Pekins, except that they possess crests and are one pound lighter in weight.

PEKIN DUCKS.

Disqualifications.

Bill or bean of drake marked with black; feathers other than white, or creamy white, in any part of plumage. (See general disqualifications.)

STANDARD WEIGHTS.

Adult Drake..........9 lbs.   Adult Duck..........8 lbs.
Young Drake..........8 lbs.   Young Duck..........7 lbs.

SHAPE OF DRAKE AND DUCK.

HEAD: Long, finely formed.

BILL: Of medium size, slightly convex between juncture with head and extremity of bill.

EYES: Large, deep set.

NECK: In drake, rather long and large; in duck, of medium length; in both, carried well forward; arched.

WINGS: Short, carried closely and smoothly against sides.

BACK: Long, broad, with a slight depression from shoulder to tail.

TAIL: Rather erect, the curled feathers of drake being hard and stiff.

BREAST: Broad, deep, prominent, carried low.

BODY: Long, broad, carried just clear of ground; carriage, slightly elevated in front, sloping downward toward rear.

LEGS AND TOES: Thighs and shanks, short, large, set well back; toes, straight, connected by web.

COLOR OF DRAKE AND DUCK.

BILL: Orange-yellow, free from black.

EYES: Deep leaden-blue.

SHANKS AND TOES: Reddish-orange.

PLUMAGE: White or creamy-white.

AYLESBURY DUCKS.

Disqualifications.

Bill or bean of drake marked with black; feathers other than pure white in any part of plumage. (See general disqualifications.)

STANDARD WEIGHTS.

Adult Drake	9 lbs.	Adult Duck	8 lbs.
Young Drake	8 lbs.	Young Duck	7 lbs.

SHAPE OF DRAKE AND DUCK.

HEAD: Large, long, finely formed.

BILL: Long, broad, outline nearly straight from top of head to tip of bill.

EYES: Full, deep set.

NECK: Long, moderately thick, slightly curved.

WINGS: Strong, carried closely and smoothly against sides.

BACK: Long, broad, straight on top.

TAIL: Only slightly elevated; composed of stiff, hard feathers; sex feathers of drake, hard, well curled.

BREAST: Deep, prominent.

BODY: Long, deep, broad; keel, straight; carriage, nearly horizontal.

LEGS AND TOES: Thighs, short, stout; shanks, strong; toes, straight, connected by web.

COLOR OF DRAKE AND DUCK.

BILL: Pale flesh-color, free from dark or black marks.

EYES: Deep leaden.

SHANKS AND TOES: Bright, light orange.

PLUMAGE: Web, quill and fluff of feathers in all sections, pure white.

ROUEN DUCKS.

Disqualifications.

Bills, clear yellow, dark green, blue or lead color; any approach to white ring on neck of duck; white in primaries or secondaries. (See general disqualifications.)

STANDARD WEIGHTS.

Adult Drake..........9 lbs.   Adult Duck...........8 lbs.
Young Drake..........8 lbs.   Young Duck..........7 lbs.

SHAPE OF DRAKE AND DUCK.

HEAD: Full and round.

BILL: Long, broad, wider at extremity than at base; top slightly depressed from crown of head to tip of bill.

EYES: Bold, full.

NECK: Long, tapering, curved, erect.

WINGS: Short, carried smoothly against sides.

BACK: Long, broad, slightly arched.

TAIL: Only slightly elevated; composed of hard, stiff feathers; sex feathers of drake, hard, well curled.

BREAST: Broad, deep.

BODY: Long, deep, broad; keel, long, straight; carriage, nearly horizontal.

LEGS AND TOES: Thighs, short, large; shanks, short, large; toes, straight, connected by web.

PEKIN DRAKE

326

PEKIN DUCK

ROUEN DRAKE

ROUEN DUCK

ROUEN DRAKE

ROUEN DUCK

COLOR OF DRAKE.

HEAD: Plumage, rich, lustrous green.

BILL: Greenish-yellow, without any other shade, except black bean at tip.

EYES: Dark brown.

NECK: Rich, lustrous green, with a distinct white ring on lower part, not quite meeting at back.

WINGS: Flights, slaty-black and brown, free from white; coverts, pale, clear gray; small coverts, gray, closely penciled; pinion coverts, dark gray to slaty-black; bars formed by a line of white in center of small coverts, which should be gray tipped with black, coming to a line at base of flight coverts, which should be slaty-black above the quill, and a rich, iridescent blue below, tipped with white at lower side, making two distinct white bars (the pinion bar being edged with black), with a distinct blue, ribbon-like mark between them.

BACK: Upper part, ashy-gray, mixed with green, becoming a rich, lustrous green on lower part of rump; shoulders, gray, finely streaked with wavy brown lines.

TAIL: Dark, ashy-brown, outer web in old birds edged with white; coverts, black, showing very rich purple reflections; tail, well supplied on outer side with solid, beetle-green feathers.

BREAST: Very rich purplish-brown or claret, extending well down on breast and free from any other color.

BODY: Upper part, steel-gray; sides, steel-gray, very finely penciled across the feathers with glossy black, growing wider near the vent and ending in a solid, greenish-black, forming a distinct line of separation between the two colors.

LEGS AND TOES: Thighs, ashy-gray; shanks and toes, orange, with brownish tinge.

COLOR OF DUCK.

HEAD: Plumage, deep brown with two light tan stripes on each side, running from bill to point behind eyes.

BILL: Brownish-orange, with dark blue blotch on upper part and black bean at tip.

EYES: Dark brown.

NECK: Golden brown, penciled with dark, lustrous brown, free from any appearance of white ring.

WINGS: Flat of wing, light brown, with distinct pencilings of rich greenish-brown; wing-bar, rich, brilliant purple, each end of bar banded with white; secondaries, dark brown, with distinct pencilings that conform to shape of feather.

BACK: Rich golden brown, richly marked with wide pencilings of greenish-black; shoulder-coverts, dark brown with distinct pencilings of light brown that conform to shape of feather.

TAIL: Golden brown, with a distinct broad, wavy penciling of dark greenish-brown; coverts, brown, with broad, distinct and regular pencilings of greenish-brown.

BREAST: Rich, golden brown, with distinct wide pencilings of light brown that conform to shape of feather.

BODY: Under part, light brown, each feather distinctly penciled with rich, dark brown to point of tail; sides, dark brown, with distinct pencilings of light brown that conform to shape of feather.

LEGS AND TOES: Thighs, dark brown, distinctly penciled; shanks and toes, orange or orange-brown.

CAYUGA DUCKS.

Disqualifications.

White in any part of plumage. (See general disqualifications.)

STANDARD WEIGHTS.

Adult Drake	8 lbs.	Adult Duck	7 lbs.
Young Drake	7 lbs.	Young Duck	6 lbs.

SHAPE OF DRAKE AND DUCK.

HEAD: Long, finely formed.

BILL: Long, top line slightly depressed.

EYES: Full.

NECK: Of medium length, slightly arched.

WINGS: Short, folded closely and smoothly against sides.

BACK: Long, broad.

TAIL: Only slightly elevated; composed of hard, stiff feathers: sex feathers of drake, hard, well curled.

BREAST: Broad, full, prominent.

BODY: Long, broad, deep; carriage, nearly horizontal.

LEGS AND TOES: Thighs, short, large; shanks, of medium length and size; toes, straight, connected by web.

COLOR OF DRAKE AND DUCK.

BILL: Black.

EYES: Dark brown.

SHANKS AND TOES: Black or dark slate.

PLUMAGE: Lustrous greenish-black throughout.

CALL DUCKS.

SHAPE OF DRAKE AND DUCK.

HEAD: Small, slender.

BILL: Short, trim.

EYES: Of medium size.

NECK: Of medium length.

WINGS: Neat, closely folded.

BACK: Comparatively short.

TAIL: Only slightly elevated; composed of hard, stiff feathers; sex feathers of drake, well curled.

BREAST: Round, full.

BODY: Short, compact, small—the smaller, the better; carriage, nearly horizontal, possessing a startled and gamy appearance.

LEGS AND TOES: Thighs, short, plump; shanks, short; toes, straight, connected by web.

GRAY CALL DUCKS.

Disqualifications.

Any approach to white ring on neck of duck; white primaries in either sex. (See general disqualifications.)

COLOR OF DRAKE.

HEAD: Plumage rich, lustrous green.

BILL: Greenish-yellow.

EYES: Dark brown.

NECK: Lustrous green with a distinct white ring on lower part, not quite meeting in back.

WINGS: Grayish-brown, mixed with green, with broad, ribbon-like mark of rich purple, showing metallic reflections of green and blue, edged with white, the two colors quite distinct; primaries, dark dusky brown.

BACK: Ashy-gray, mixed with green on upper part; on lower part and rump, rich, lustrous green.

TAIL: Dark, ashy-brown; outer web in old birds, edged with white; tail-coverts, black, showing very rich purple reflections.

BREAST: Rich purplish-brown or claret, extending well down on breast and free from any other color.

BODY: Under part and sides, steel gray, growing lighter near vent and ending in solid, beetle-green black, forming distinct line of separation between the two colors.

LEGS AND TOES: Thighs, ashy-gray; shanks and toes, orange, with a brownish tinge.

COLOR OF DUCK.

HEAD: Plumage, deep brown, with two light tan stripes on each side, running from bill to a point behind eyes.

BILL: Brownish-orange.

EYES: Dark brown.

NECK: Light brown, penciled with dark, lustrous brown, free from any appearance of a white ring.

WINGS: Light brown, mixed with green, with broad, ribbon-like bars of rich purple across them, edged with white, the two colors distinct; primaries, brown.

BACK: Light brown, richly marked with green.

TAIL: Light brown, with distinct, broad, wavy penciling of dark greenish-brown; tail coverts, brown, with broad, distinct and regular pencilings of dark brown or greenish-brown.

BREAST: Dark brown, richly penciled with a lighter brown.

BODY: Under part and sides, light brown, each feather distinctly penciled with rich, dark brown to point of tail.

LEGS AND TOES: Thighs, dark brown, distinctly penciled; shanks and toes, orange, or orange-brown.

WHITE CALL DUCKS.

Disqualifications.

Bill or bean of drake marked with black; feathers other than white or creamy in any part of the plumage. (See general disqualifications.)

COLOR OF DRAKE AND DUCK.

BILL: Bright yellow.

EYES: Blue.

SHANKS AND TOES: Bright orange.

PLUMAGE: Web, quill and fluff of feathers in all sections, pure white.

333

BLACK EAST INDIA DUCKS.

Disqualifications.

White in any part of plumage. (See general disqualifications.)

SHAPE OF DRAKE AND DUCK.

HEAD: Short.

BILL: Rather short.

EYES: Of medium size.

NECK: Short, nicely arched.

WINGS: Long, well folded.

BACK: Of medium width, rather long.

TAIL: Slightly elevated, composed of short, stiff feathers; sex feathers of drake, well curled.

BREAST: Full, plump.

BODY: Long, comparatively small—the smaller, the better; carriage, nearly horizontal, possessing a gamy appearance.

LEGS AND TOES: Thighs, short, plump; shanks, short; toes, straight, connected by web.

COLOR OF DRAKE AND DUCK.

BILL: Drake, very dark green; duck, black.

EYES: Dark brown.

SHANKS AND TOES: Black.

PLUMAGE: Rich black, with brilliant, greenish tint.

MUSCOVY DUCKS.

STANDARD WEIGHTS.

Adult Drake	10 lbs.	Adult Duck	7 lbs.
Young Drake	8 lbs.	Young Duck	6 lbs.

SHAPE OF DRAKE AND DUCK.

HEAD: Rather long; in drake, large, the top covered with long, crest-like feathers, which are elevated or depressed by the bird when it becomes excited or alarmed; sides of head and face covered with caruncles—the larger, the better.

BILL: Rather short, of medium width.

EYES: Of medium size, having slightly over-arched socket.

NECK: Of medium length, well arched.

WINGS: Very long, stout.

BACK: Long, broad, somewhat flat.
TAIL: Rather long, with abundance of stiff plumage.
BREAST: Broad, full.
BODY: Long, broad; carriage, nearly horizontal.
LEGS AND TOES: Thighs, very short, large; shanks, short, large; toes, straight, connected by web.

COLORED MUSCOVY DUCKS.

Disqualifications.

Smooth heads; plumage more than one-half white. (See general disqualifications.)

COLOR OF DRAKE AND DUCK.

HEAD: Plumage, glossy black and white.
BILL: Pink, shaded with horn.
EYES: Brown.
FACE: Caruncles, red.
NECK: Black, or black and white, black predominating.
WINGS: Coverts, rich, lustrous greenish-black.
BACK: Lustrous blue-black, sometimes broken with white feathers.
TAIL: Black.
BREAST AND BODY: Lustrous blue-black, sometimes broken with white; the blacker the plumage, the better.
LEGS AND TOES: Thighs, white or black, white preferred; shanks and toes, varying from yellow to dark lead.

WHITE MUSCOVY DUCKS.

Disqualifications.

Smooth heads; feathers other than pure white in any part of plumage. (See general disqualifications.)

COLOR OF DRAKE AND DUCK.

BILL: Pinkish flesh-color.
EYES: Blue.
FACE: Caruncles, red.
SHANKS AND TOES: Pale orange or yellow.
PLUMAGE: Web, quill and fluff of feathers in all sections, pure white.

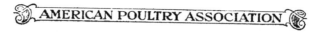

BLUE SWEDISH DUCKS.

Disqualifications.

Yellow bills; absence of white in breast; feathers of any other color than blue forming more than one-fourth of plumage. (See general disqualifications.)

STANDARD WEIGHTS.

Adult Drake ....8 lbs.   Adult Duck.........7 lbs.
Young Drake.......6½ lbs.   Young Duck........5½ lbs.

SHAPE OF DRAKE AND DUCK.

HEAD: Long, finely formed.

BILL: Of medum size; nearly straight in outline when viewed sidewise.

EYES: Full.

NECK: Long, slightly arched.

WINGS: Short, carried closely.

BACK: Long, broad, with slight concave sweep from shoulders to tail.

TAIL: Slightly elevated; sex feathers of drake, well curled.

BREAST: Full, deep.

BODY: Broad, of medium length, rangy; carriage, nearly horizontal, somewhat elevated in front.

LEGS AND TOES: Thighs, short, stout; shanks, stout; toes, straight, connected by web.

COLOR OF DRAKE AND DUCK.

HEAD: Drake, dark blue, sometimes approaching black, with a green sheen; duck, same as general body color.

BILL: Drake, greenish-blue; duck, smutty brown, with a dark brown blotch, similar to Rouen Duck blotch, only larger.

EYES: Dark brown.

WINGS: Two flight feathers, pure white; balance of wing, uniform with general plumage.

BREAST: Front part, pure white, forming heart-shaped spot about three by four inches in size, often extending upward to lower mandible.

SHANKS AND TOES: Reddish-brown or grayish-black, the former preferred.

PLUMAGE: Uniform steel-blue throughout, except as noted above.

CRESTED WHITE DUCKS.

SCALE OF POINTS.

Symmetry	4
Weight	6
Condition	10
Head — Shape 2, Color 2	4
Bill — Shape 2, Color 2	4
Eye — Shape 2, Color 2	4
Crest	15
Neck — Shape 4, Color 3	7
Wings — Shape 4, Color 6	10
Back — Shape 6, Color 4	10
Tail — Shape 2, Color 2	4
Breast — Shape 6, Color 4	10
Body — Shape 6, Color 2	8
Legs and Toes — Shape 2, Color 2	4
	100

Disqualifications.

Absence of crest, or crest falling over to one side; bills and legs other than yellow; feathers other than white or creamy white in any part of plumage. (See general disqualifications.)

STANDARD WEIGHTS.

Adult Drake	7 lbs.	Adult Duck	6 lbs.
Young Drake	6 lbs.	Young Duck	5 lbs.

SHAPE OF DRAKE AND DUCK.

HEAD: Of medium size.

BILL: Of medium size.

EYES: Large.

CREST: Large, well balanced on crown of head.

NECK: Rather long; slightly arched.

WINGS: Of medium length; smoothly folded.

BACK: Of medium length and width.

TAIL: Only slightly elevated; composed of hard, stiff feathers; sex feathers of drake, well curled.

BREAST: Prominent, full.

BODY: Of medium length, plump; carriage, nearly horizontal.

LEGS AND TOES: Thighs, short, plump; shanks, short; toes, straight, connected by web.

COLOR OF DRAKE AND DUCK.

BILL: Yellow.

EYES: Blue.

SHANKS AND TOES: Light orange.

PLUMAGE: Web, quill and fluff of feathers in all sections, pure white.

BUFF DUCKS.

Disqualifications.

Color of plumage other than buff or seal-brown. (See general disqualifications.)

STANDARD WEIGHTS.

Adult Drake ....8 lbs.   Adult Duck ....7 lbs.
Young Drake ....7 lbs.   Young Duck ....6 lbs.

SHAPE OF DRAKE AND DUCK.

HEAD: Oval, fine, racy.

BILL: Moderate in length, straight in line from skull.

EYES: Bold, full.

NECK: Fairly long and gracefully curved.

WINGS: Short; carried closely and smoothly against side.

BACK: Broad, long.

TAIL: Small, rising gently; sex feathers of drake, well curled.

BREAST: Broad, deep, prominent, carried moderately low.

BODY: Long, broad, deep.

LEGS AND TOES: Thighs and shanks, short, large, set well apart; toes, straight, connected by web.

COLOR OF DRAKE AND DUCK.

BILL: Yellow in drake, brownish orange in duck with dark bean.

EYES: Brown, with blue pupil.

SHANKS AND TOES: Orange yellow.

PLUMAGE: Surface throughout an even shade of rich fawn buff with the exception of head and upper portion of neck in drake which should be seal-brown. Penciling to be considered a serious defect.

338

RUNNER DUCKS.

SCALE OF POINTS.

Symmetry	4
Carriage	15
Weight	6
Condition	6
Head — Shape 3, Color 4	7
Bill — Shape 5, Color 2	7
Eyes	2
Neck — Shape 8, Color 4	12
Wings — Shape 3, Color 3	6
Back — Shape 6, Color 3	9
Tail — Shape 2, Color 2	4
Breast — Shape 6, Color 3	9
Body — Shape 6, Color 3	9
Legs and Toes — Shape 2, Color 2	4
	100

STANDARD WEIGHTS.

Adult Drake .... 4½ lbs.   Adult Duck 4   lbs.
Young Drake 4   lbs.   Young Duck .... 3½ lbs.

SHAPE OF DRAKE AND DUCK.

HEAD: Long, flat, finely formed.

BILL: Strong at base, fairly broad and long, extending down from the skull in a straight line, giving it the appearance of a long wedge.

EYES: Set high in head.

NECK: Long, thin, line of neck almost straight.

WINGS: On medium length, carried closely to body.

BACK: Long, straight, narrow.

TAIL: Composed of hard, stiff feathers; sex feathers of drake, hard, well curled.

BREAST: Full, but not rounded, showing but little suggestion of keel, carried well up.

BODY: Long, narrow, racy-looking; carried erectly with no indication of keel; carriage, very erect.

LEGS AND TOES: Legs, of medium length; toes, straight, connected by web.

RUNNER DRAKE

RUNNER DUCK

341

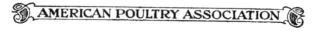

FAWN AND WHITE RUNNER.

Disqualifications.

Claret breast; blue wing-bars; absence of two or more primaries or secondaries; scoop bill. (See general disqualifications.)

COLOR OF DRAKE AND DUCK.

HEAD: Fawn and white. A line of white divides the cap from the cheek markings behind the eyes, and a narrow line of white divides the base of the bill from the head markings.

BILL: Drake, yellow when young; greenish-yellow when fully developed. Duck, yellow spotted with green when young; a dull green when fully matured, with black bean in both duck and drake.

EYES: Dark brown.

NECK: The upper two-thirds white, lower one-third fawn, sharply defined.

WINGS: Wing-bows and coverts, fawn; primaries and secondaries, white.

BACK: Even fawn throughout.

TAIL: Fawn.

BREAST: Fawn and white, divided about halfway between point of breast-bone and legs; upper half, fawn; lower part, white.

BODY AND FLUFF: White, except an indistinct line of fawn, which runs from base of tail to thigh.

SHANKS AND TOES: Orange red.

WHITE RUNNER

Disqualifications.

Black bean in drake, color other than white or creamy white in any part of plumage; absence of two or more primaries or secondaries; scoop bill. (See general disqualifications.)

COLOR OF DRAKE AND DUCK.

HEAD: White.

BILL: Yellow.

EYES: Leaden-blue.

SHANKS AND TOES: Orange.

PLUMAGE: Web, quill and fluff of feathers in all sections, pure white.

PENCILED RUNNER

Disqualifications.

Blue wing-bar; claret breast; bright green head like Rouen or Mallard; absence of two or more primaries or secondaries; scoop bill. (See general disqualifications.)

COLOR OF DRAKE.

HEAD: Dull bronze green, a narrow line of white extending to and encircling the eyes, dividing cap and cheek markings; head markings and bill divided by a narrow line of white; all markings clear cut.

BILL: Of young specimens, yellow; of matured, greenish-yellow; bean, black.

EYES: Dark brown.

NECK: Upper two-thirds, white, lower one-third medium fawn, sharply defined.

WINGS: Shoulders and top part of wings, fawn, same shade as breast; primaries and secondaries, white, the white extending up the lower part of wings to a point a little above the lower part of the body, forming an inverted V-shaped marking on the side of the body; the color of shoulders and top part of wings when folded come to a point in lower part of back, forming a heart-shape, like a heart pressed on the back.

BACK: Medium fawn, the feathers when examined closely show a soft fawn ground finely stippled with a slightly darker shade of fawn.

TAIL: Dull bronze-green.

BREAST: Medium fawn and white evenly divided about half way between point of breast bone and the legs; upper section dark fawn; lower section white.

BODY AND FLUFF: Body, medium fawn, same shade as breast; fluff white, except an indistinct line of fawn which runs from the base of tail to thighs.

SHANKS AND TOES: Orange-red.

COLOR OF DUCK.

HEAD: Medium fawn, a line of white divides the cap and cheek markings and a narrow line of white divides the base of bill from head markings.

BILL: Spotted with green when young; dull green when matured; bean, black.

EYES: Dark brown.

PLUMAGE: White markings same as drake, colored markings medium fawn throughout, surface color should be even, no fawn colored section being lighter or darker than another. The under-color a medium or darker shade of fawn, a light line of fawn color running around near the edge of each feather, the border or edge a darker shade. The penciling may be more prominent on the back and wings.

SHANKS AND TOES: Orange-red.

GEESE.

Breeds	*Varieties*
TOULOUSE	Gray
EMBDEN	White
AFRICAN	Gray
CHINESE	{ Brown / White
WILD OR CANADIAN	Gray
EGYPTIAN	Colored

SCALE OF POINTS.

Symmetry	4
Weight	6
Condition	10
Head—Shape 6, Color 4	10
Eye—Shape 2, Color 2	4
Neck—Shape 6, Color 3	9
Wings—Shape 6, Color 6	12
Back—Shape 8, Color 4	12
Tail—Shape 2, Color 2	4
Breast—Shape 8, Color 4	12
Body—Shape 10, Color 4	14
Legs and Toes—Shape 2, Color 1	3
	100

GEESE.

The male of the Toulouse, Embden and African varieties—the heavy weights of the goose family—should be broad and flat in back, with deep, round, full breast and long body,—these features giving the bird a massive appearance. The head should be large, the neck of good length, and slightly arched, and the bird well balanced in carriage. The female should resemble the male, except that she is somewhat less massive.

Chinese Geese are an ornamental variety of medium size, with long, arched necks, carried very upright, and having a large knob at base of beak. In body they are short and upright, the general effect being novel and striking.

Wild or Canadian Geese, now domesticated, are seen most frequently in public parks. They are of medium size, with long arched, snake-like necks, and have small heads, carried well elevated, which gives them a bold and defiant appearance.

Egyptian Geese are decidedly odd members of the goose family, being very different in most respects from the other standard varieties. Some writers have gone so far as to separate them from the goose tribe. Striking characteristics are their small and pugnacious disposition.

TOULOUSE.

Disqualifications.

White feathers in primaries or secondaries. (See general disqualifications.)

STANDARD WEIGHTS.

Adult Gander ....26 lbs.	Adult Goose ....20 lbs.
Young Gander ......20 lbs.	Young Goose ....16 lbs.

SHAPE OF GANDER AND GOOSE.

HEAD: Rather large, short.

BILL: Comparatively short, stout at base.

EYES: Large, full.

NECK: Of medium length, carried rather erect; dewlap very desirable in aged fowls.

346

WINGS: Large, strong, smoothly folded against sides.

BACK: Of moderate length, broad, curving slightly from neck to tail.

TAIL: Comparatively short; feathers, hard, stiff.

BREAST: Broad, deep.

BODY: Of good length, broad, very deep, compact; in fat specimens, almost touching the ground; keel, deep with straight line from breast to abdomen; stern, almost square.

LEGS AND TOES: Thighs and shanks, short, stout; toes, straight, connected by web.

COLOR OF GANDER AND GOOSE.

HEAD: Gray.

BILL: Pale orange.

EYES: Dark brown or hazel.

NECK: Dark blue-gray, shading to lighter gray as it approaches the black.

WINGS: Primaries, dark gray; secondaries, darker than primaries, with very narrow edging of lighter gray; coverts, dark gray, with very narrow edging of lighter gray.

BACK: Dark gray.

TAIL: Gray and white, the ends tipped with white.

BREAST: Light gray, edged with white.

BODY: Underneath, light gray, growing lighter until it becomes almost white on abdomen, the white extending back to and around tail; sides, light gray, becoming dark blue-gray over thighs, edged with lighter gray; white covering all lower posterior parts; from lower front view very little white visible.

LEGS AND TOES: Thighs, light gray; shanks and toes, deep reddish-orange.

EMBDEN GEESE.

Disqualifications.

Feathers other than white in any part of plumage, except traces of gray in wings and backs of young specimens. (See general disqualifications.)

STANDARD WEIGHTS.

Adult Gander.........20 lbs.	Adult Goose..........18 lbs.
Young Gander........18 lbs.	Young Goose.........16 lbs.

TOULOUSE GANDER

TOULOUSE GOOSE

SHAPE OF GANDER AND GOOSE.

HEAD: Rather large.

BILL: Of medium length and size; stout at base.

EYES: Large.

NECK: Rather long, carried quite upright.

WINGS: Large, well rounded, strong, smoothly folded against sides.

BACK: Long and straight.

TAIL: Comparatively short; feathers, hard and stiff.

BREAST: Round, deep, full, without keel.

BODY: Large, square, very deep; in fat specimens almost touching the ground; abdomen, full and deep.

LEGS AND TOES: Thighs, short, large; shanks, short, stout; toes, straight, connected by web.

COLOR OF GANDER AND GOOSE.

BILL: Orange.

EYES: Bright blue.

SHANKS AND TOES: Deep orange.

PLUMAGE: Pure white.

AFRICAN GEESE.

Disqualifications.

Bill and knob other than black; absence of dewlap in adult specimens; white feathers in primaries and secondaries. (See general disqualifications.)

STANDARD WEIGHTS.

Adult Gander ....20 lbs. Adult Goose ....18 lbs.
Young Gander ....16 lbs. Young Goose ....14 lbs.

SHAPE OF GANDER AND GOOSE.

HEAD: Large, with large knob, heavy dewlap under throat, which in young birds is but slightly developed.

KNOB: Large.

BILL: Rather large, stout at base.

EYES: Large.

NECK: Long, curved; throat embellished with dewlap.

WINGS: Large, strong, smoothly folded against sides.

BACK: Broad, flat.

TAIL: Composed of stiff, hard feathers.
BREAST: Round, moderately full.
BODY: Large, long, carried rather upright.
LEGS AND TOES: Thighs, short, stout; shanks, of medium length; toes, straight, connected by web.

COLOR OF GANDER AND GOOSE.

HEAD: Black or very dark gray.
KNOB: Black.
BILL: Black.
EYES: Dark brown.
NECK: Light gray, with a dark gray stripe down back of neck from head to body.
WINGS: Dark gray.
BACK: Dark gray.
TAIL: Dark gray.
BREAST: Gray.
BODY: Light gray on under parts.
LEGS AND TOES: Thighs, light gray; shanks and toes, dark orange.

CHINESE GEESE.

STANDARD WEIGHTS.

Adult Gander ....12 lbs.	Adult Goose ....10 lbs.
Young Gander ....10 lbs.	Young Goose .... 8 lbs.

SHAPE OF GANDER AND GOOSE.

HEAD: Of medium size, with large knob at base of bill.
KNOB: Large, the larger the better.
BILL: Of medium length, stout at base.
EYES: Large.
NECK: Long, gracefully arched, carried very upright.
WINGS: Large, strong, smoothly folded against sides.
BACK: Of medium length and width, slightly arched from neck to tail.
TAIL: Composed of hard stiff feathers.
BREAST: Round, full.
BODY: Rather short, round, plump.
LEGS AND TOES: Thighs, short, stout; shanks, of medium length; toes, straight, connected by web.

351

WHITE CHINESE GANDER

WHITE CHINESE GOOSE

BROWN CHINESE.

Disqualifications.

Absence of knob; white feathers in primaries or secondaries. (See general disqualifications.)

COLOR OF GANDER AND GOOSE.

HEAD: Brown.

KNOB: Dark brown or black.

BILL: Black.

EYES: Hazel or brown.

NECK: Light brown or grayish-brown, with a dull yellowish-brown stripe down back of neck from head to body.

WINGS: Grayish-brown.

BACK: Dark brown.

TAIL: Grayish-brown.

BREAST: Grayish-brown.

BODY: Grayish-brown, lighter shade on under parts.

LEGS AND TOES: Thighs, grayish-brown; shanks and toes, dusky orange.

WHITE CHINESE.

Disqualifications.

Absence of knob; feathers other than pure white in any part of plumage. (See general disqualifications.)

COLOR OF GANDER AND GOOSE.

KNOB: Orange.

BILL: Orange.

EYES: Light blue.

SHANKS AND TOES: Orange-yellow.

PLUMAGE: Pure white.

WILD OR CANADIAN GEESE.

Disqualifications.

The clipping of one wing to prevent flying is not to handicap specimen. (See general disqualifications.)

STANDARD WEIGHTS.

Adult Gander ....12 lbs.   Adult Goose ....10 lbs.
Young Gander ......10 lbs.   Young Goose .... 8 lbs.

SHAPE OF GANDER AND GOOSE.

HEAD: Rather small.
BILL: Small, tapering toward point.
EYES: Prominent, sharp, bold.
NECK: Long, slender, snaky in appearance.
WINGS: Long, large, powerful.
BACK: Long, rather narrow, arched from neck to tail.
TAIL: Composed of hard, stiff feathers.
BREAST: Full, deep.
BODY: Rather long, somewhat slender.
LEGS AND TOES: Thighs, rather short; shanks, rather long; toes, straight, of medium length, connected by web.

COLOR OF GANDER AND GOOSE.

HEAD: Black, with a white stripe nearly covering side of face.
BILL: Black.
EYES: Black.
NECK: Black, shading to light gray at base.
WINGS: Dark gray; primaries, dusky-black, showing only dark gray when wing is folded; secondaries, brown, lighter than primaries.
BACK: Dark gray.
TAIL: Glossy black.
BREAST: Light gray, growing darker as it approaches legs.
BODY: Under part of body, from legs to tail, white.
LEGS AND TOES: Thighs, gray; shanks and toes, black.

355

EGYPTIAN GEESE.

Disqualifications.

The clipping of one wing to prevent flying is not to handicap specimen. (See general disqualifications.)

STANDARD WEIGHTS.

Adult Gander .... 10 lbs.   Adult Goose .... 8 lbs.
Young Gander .... 8 lbs.   Young Goose .... 6 lbs.

SHAPE OF GANDER AND GOOSE.

HEAD: Small, rather long.
BILL: Of medium length and size.
EYES: Prominent, bold.
NECK: Of medium length, rather small.
WINGS: Large; on wing-joint, in place of the ordinary hard knobs, there are strong, white, horny spurs, about five-eighths of an inch long.
BACK: Rather narrow, slightly arched from neck to tail.
TAIL: Composed of hard, stiff feathers.
BREAST: Round, not deep.
BODY: Rather long, somewhat small and slender.
LEGS AND TOES: Thighs, of medium length, stout; shanks, rather long; toes, straight, long, connected by web.

COLOR OF GANDER AND GOOSE.

HEAD: Black and gray, with chestnut patch around eyes.
BILL: Purple or bluish-red.
EYES: Orange.
NECK: Gray and black.
WINGS: Shoulders, white with narrow black stripe or bar of rich metallic luster; primaries and secondaries, glossy black.
BACK: Gray and black.
TAIL: Glossy black.
BREAST: Center, chestnut; remainder, gray.
BODY: Gray and black on upper parts; under parts, buff or yellow, distinctly and regularly penciled with black lines.
LEGS AND TOES: Thighs, pale buff; shanks and toes, reddish-yellow.

TURKEYS.

Varieties

Turkeys..................................... {
Bronze
Narragansett
White Holland
Black
Slate
Bourbon Red
}

SCALE OF POINTS.

Symmetry	4
Weight	18
Condition	4
Head — Shape 2, Color 2	4
Eye — Shape 2, Color 2	4
Throat Wattle	4
Neck — Shape 3, Color 2	5
Wings — Shape 4, Color 6	10
Back — Shape 4, Color 6	10
Tail — Shape 4, Color 8	12
Breast — Shape 5, Color 5	10
Body and Fluff — Shape 5, Color 5	10
Legs and Toes — Shape 3, Color 2	5
	100

TURKEYS.

The turkey is evidently of American nativity. The Wild Turkey was found in great numbers by the pioneers in the Eastern, Southern and Western sections of the United States. They existed also in great numbers in Mexico. The Turkey was introduced into Europe from America. The Turkey male should be large in frame and deep in body, with a broad, round, full breast that varies in prominence according to the variety and gives the fowl a stately and majestic appearance. The head should be of good size, and the eyes possess an alert and bold expression. The legs and shanks must be large, straight and well set. The outline of all sections should be in perfect symmetry. The Turkey female should be large in frame and deep in body, with a broad, round, full breast, being similar in all sections to the male, except finer in structure. She, too, is stately in appearance. The head should be of good size, with a bright, watchful eye; the legs and shanks large, straight and well set, the outlines of each section conforming to those of the male, except in size.

SHAPE OF MALE AND FEMALE.

HEAD: Long, broad, carunculated.

BEAK: Strong, curved, well set in head.

EYES: Oval, prominent.

THROAT WATTLE: Heavily carunculated.

NECK: Long, curving backward toward tail.

BEARD: Long, bristly, prominent.

WINGS: Large, powerful, smoothly folded, carried well up on sides.

BACK: Broad, somewhat curving, rising from neck and descending in graceful curve to outer end of tail.

TAIL: Rather long.

BREAST: Broad, deep, full, well rounded, carried well forward.

BODY: Long, deep through middle, finely rounded.

LEGS AND TOES: Thighs, long, stout; shanks, large, long, strong; toes, straight, strong.

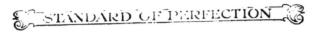

BRONZE TURKEYS.

Disqualifications.

White feathers in any part of plumage; wings showing one or more primary or secondary feathers clear black or brown, or absence of white or gray bars more than one-half of the length of primaries; color of back, tail or tail-coverts, clear black, brown or gray. (See general disqualifications.)

Note: The following defects should be cut severely: Absence of one or more primary or secondary wing feathers; absence of one or more center main tail feathers; white or gray bars, other than the terminating wide edging of white, showing on main tail feathers; absence of black bands on one or more of the large main tail-coverts; decidedly wry wings; decidedly crooked breast-bone.

STANDARD WEIGHTS.

Adult Cock	36 lbs.	Hen	20 lbs.
Yearling Cock	33 lbs.	Pullet	16 lbs.
Cockerel	25 lbs.		

When two specimens are both over Standard weight and equal in all other points, the one nearest Standard weight shall win.

COLOR OF MALE.

HEAD: Rich red, changeable to bluish-white.

BEAK: Light horn at tip, dark at base.

EYES: Dark brown.

THROAT WATTLE: Rich red, changeable to bluish-white.

NECK: Rich, brilliant copperish-bronze.

BEARD: Black.

WINGS: Bows, rich, brilliant, copperish bronze, ending in a narrow band of black; coverts, a bright, rich, copperish-bronze, forming a beautiful, broad, bronze band across wings when folded, feathers terminating in a wide, black band, forming a glossy, ribbon-like mark, which separates them from primaries and secondaries; primaries, each feather, throughout its entire length, alternately crossed with distinct, parallel black and white bars of equal width, running straight across the feathers; flight coverts, barred similar to primaries; secondaries, dull black, the more distinct the better, the color changing to a bronze-brown as the middle of the back is approached and the white bars becoming less distinct, an edging of brown in secondaries being very objectionable.

BRONZE TURKEY MALE

BRONZE TURKEY FEMALE

BACK: From neck to middle of back, a rich brilliant, copperish bronze, each feather terminating in a very narrow, black band, extending across end; from middle of back to tail-coverts black, each feather having a brilliant, copperish-bronze band extending across it near the end.

TAIL: Dull black, each feather evenly and distinctly marked transversely with parallel lines of brown; each feather having a wide black band extending across it near the end (the more bronze on this band, the better), and terminating in a wide edging of pure white. Coverts, dull black, each feather evenly and distinctly marked transversely with parallel lines of brown, each feather having a wide black and bronze band extending across it near the end, terminating in a wide edging of pure white, the few larger coverts extending well out on tail, having little bronze on them. The more distinct the colors throughout the whole plumage, the better.

BREAST: Rich, brilliant, copperish-bronze; feathers on lower part of breast, approaching the body, terminate in a black band extending across the end.

BODY AND FLUFF: Body, black, each feather with a wide, brilliant, copperish-bronze band extending across it near the end and terminating in a narrow edging of pure white; fluff, dull black.

LEGS AND TOES: Thighs, similar to breast, but less brilliant in shade; shanks and toes, in mature birds, pinkish; in young birds, dark, approaching black.

COLOR OF FEMALE.

PLUMAGE: Similar to that of male, except an edging of white on feathers of back, wing-bows, wing-coverts, breast and body, which edging should be narrow in front, gradually widening as it approaches the rear.

BEAK, EYES, THROAT WATTLE, LEGS AND TOES: Same as male.

NARRAGANSETT TURKEYS.

Disqualifications.

Buff or slate-colored feathers in any part of plumage. (See general disqualifications.)

STANDARD WEIGHTS.

Adult Cock	30 lbs.	Hen	18 lbs.
Yearling Cock	25 lbs.	Pullet	12 lbs.
Cockerel	20 lbs.		

COLOR OF MALE.

HEAD: Rich red, changeable to bluish-white.

BEAK: Light horn.

EYES: Brown.

THROAT WATTLE: Rich red, changeable to bluish-white.

NECK: Upper part, black, each feather ending in a broad, steel-gray band; lower part, black, each feather ending in a broad, steel-gray band, edged with black, the edging of the black increasing as the body is approached.

BEARD: Black.

WINGS: Bows, metallic black, each feather ending with a band of steel-gray, edged with metallic black; coverts, inside webs, black or brown, outside webs a light steel-gray, approaching white, terminating in a metallic black band, forming a wide, steel-gray band across wing when folded; primaries, each feather evenly and distinctly barred across with parallel bars of black and white; secondaries, marked similar to primaries, but less distinct and approaching a light gray.

BACK: Rich, metallic black; saddle, black, each feather ending in a broad, steel-gray band approaching white, the light band increasing as the tail-coverts are approached.

TAIL: Dull black, each feather regularly penciled with parallel lines of light brown, ending in a broad band of metallic black, edged with steel-gray approaching white; coverts, dull black, each feather regularly penciled with parallel lines of light brown, ending in a band of metallic black, with a wide edging of light steel-gray approaching white.

BREAST: Metallic black, each feather ending in a broad, light, steel-gray band edged with black.

BODY AND FLUFF: Body, metallic black, each feather ending in a broad, light steel-gray band edged with black.

LEGS AND TOES: Thighs, rich, metallic black, each feather ending in a light steel-gray band edged with black; shanks and toes, deep salmon.

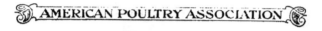

COLOR OF FEMALE.

PLUMAGE: Similar to that of male, except back shall be metallic black, each feather ending in a broad band of light steel-gray approaching white; the color of the other sections being not so distinct as in the male, and the feathers terminating in an edging of light gray approaching white.

WHITE HOLLAND TURKEYS.

Disqualifications.

Feathers other than white in any part of plumage; color of legs other than white or pinkish-white. (See general disqualifications.)

STANDARD WEIGHTS.

Adult Cock	28 lbs.	Hen	18 lbs.
Yearling Cock	24 lbs.	Pullet	14 lbs.
Cockerel	20 lbs.		

COLOR OF MALE AND FEMALE.

HEAD: Rich red, changeable to bluish-white.
BEAK: Pinkish or flesh.
EYES: Dark brown.
THROAT WATTLE: Rich red, changeable to bluish-white.
SHANKS AND TOES: White or pinkish-white.
PLUMAGE: Web, quill and fluff of feathers in all sections, pure white, except beard in male, which is deep black.

BLACK TURKEYS.

Disqualifications.

Feathers other than black in any part of plumage. (See general disqualifications.)

STANDARD WEIGHTS.

Adult Cock	27 lbs.	Hen	18 lbs.
Yearling Cock	22 lbs.	Pullet	12 lbs.
Cockerel	18 lbs.		

COLOR OF MALE AND FEMALE.

HEAD: Rich red, changeable to bluish-white.
BEAK: Dark horn, or slaty-black.
EYES: Dark brown.
THROAT WATTLE: Rich red, changeable to bluish-white.
SHANKS AND TOES: Dark lead, or slaty-black.
PLUMAGE: Surface, lustrous greenish-black throughout.

SLATE TURKEYS.

Disqualifications.

Feathers other than slaty or ashy-blue, which may be dotted with black, in any part of plumage. (See general disqualifications.)

STANDARD WEIGHTS.

Adult Cock	27 lbs.	Hen	18 lbs.
Yearling Cock	22 lbs.	Pullet	12 lbs.
Cockerel	18 lbs.		

COLOR OF MALE AND FEMALE.

HEAD: Rich red, changeable to bluish-white.
BEAK: Light blue, dark blue, or horn.
EYES: Dark hazel.
THROAT WATTLE: Rich red, changeable to bluish-white.
SHANKS AND TOES: Light or dark blue.
PLUMAGE: Slaty or ashy-blue, sometimes dotted with black, but the freer from dotting the better.

BOURBON RED TURKEY MALE

BOURBON RED TURKEY FEMALE

BOURBON RED TURKEYS.

Disqualifications.

More than one-fourth any other color than white showing in either primaries, secondaries or main tail feathers. (See general disqualifications.)

STANDARD WEIGHTS.

Adult Cock .... 30 lbs.   Hen .... 18 lbs.
Yearling Cock .... 25 lbs.   Pullet .... 12 lbs.
Cockerel .... 20 lbs.

COLOR OF MALE.

HEAD: Rich red, changeable to bluish-white.
BEAK: Light horn at tip, darker at base.
EYES: Dark brown.
THROAT WATTLE: Rich red, changeable to bluish-white.
NECK: Deep brownish-red.
WINGS: Bows, deep brownish-red; primaries and secondaries, white.
BACK: Deep brownish-red.
TAIL: White.
BREAST: Deep brownish-red.
BODY AND FLUFF: Body, deep brownish-red; fluff, brownish-red.
LEGS AND TOES: Thighs, deep brownish-red; shanks and toes, reddish-pink.

COLOR OF FEMALE.

Similar to that of male, except a narrow edging of white on breast, body and thighs.